GERMAN BATTLETANKS IN COLOR 1934-45

"NEWLY BUILT VEHICLE"-PANZER I-PANZER II-PANZER III-PANZER IV-PANZER V "PANTHER"-PANZER VI "TIGER" and "KING TIGER"-"MAUS"

HORST SCHEIBERT

1469 Morstein Road, West Chester, Pennsylvania 19380

Originally published under the title, "Deutsche Kampfpanzer in Farbe." 1934-35, copyright Podzun-Pallas-Verlag, 6360 Freidberg 3, Markt 9, West Germany, © 1985, ISBN: 3-7909-0239-X.

Translated from the German by Dr. Edward Force.

Library of Congress Catalog Number: 89-063370.

Printed in the United States of America.
ISBN: 0-88740-208-9

Sources:

—Podzun-Pallas Publishers Archives
—H. L. Doyle (drawings)
—Color pictures: Horst Helmus

Foreword

Before the war they were called battle tanks, armored (Panzer) battle tanks, or simply Panzer. In order to differentiate them from the armored vehicles for other purposes, which appeared in numerous versions during the course of the war, a more precise term was needed. The tanks leading an attack were now called battle tanks (Kampfpanzer).

The task of this special volume is to present in concentrated form all German battle tanks of the Wehrmacht and show their development, dependence and thus relationships. Mention will also be made briefly of their different versions and special types. The presentation will be completed with drawings, photos and technical data on the main versions. The latter will apply to the most powerful (usually the last) versions and show only the most important data for one type. For the horsepower listing, the sustained performances have been used, since the highest power is only of a theoretical kind. The comparison of data on page 52 is of particular interest.

The Battle Tank "Neubaufahrzeug" (Newly Built Vehicle)

Experience in the building of heavy armored vehicles was gained between 1927 and 1929 under the disguised name of "Heavy Tractor". But only in 1933 could the Army High Command grant a contract for the development of this tank, including not only a chassis and motor but also armor and weapons, under the name of "Newly Built Vehicle". The firms of Rheinmetall and Krupp each designed a version. The designs differed mainly in the arrangement of their turret armament. Rheinmetall had a 3.7 cm tank cannon (KwK) installed over a 7.5 cm KwK; Krupp had a 3.7 cm KwK beside a 10.5 cm KwK. Both of them, according to contract, showed two additional smaller turrets, each armed with two machine guns, diagonally positioned on the hull before and behind the main turret. In this way these vehicles strongly resembled the English "Independent" and Russian T-32 tanks. Five prototypes had been built by 1935, but then the heavy tank project was cancelled in favor of the Panzer IV medium tank. Only in 1942 did the development of a new heavy tank begin, which led to the production of the Tiger I and II.

Three of the five prototypes built—in fact, those with the Krupp turret—saw service in Norway in 1940, meant to give the appearance of having heavy tanks on hand there. One was lost there; the two others are said to have seen service in Russia in 1941.

Battle Tank "Neubaufahrzeug"

DESIGN OF THE KRUPP FIRM

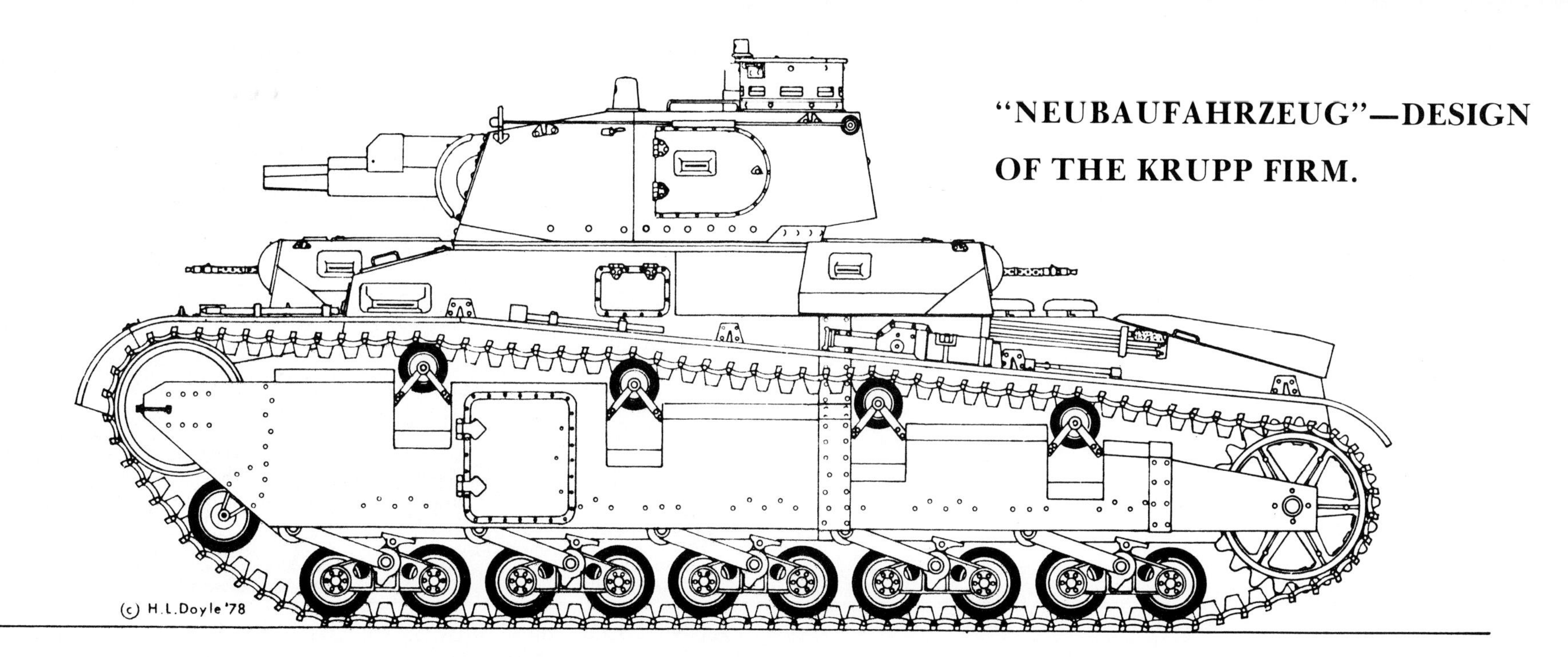

"NEUBAUFAHRZEUG"—DESIGN OF THE KRUPP FIRM.

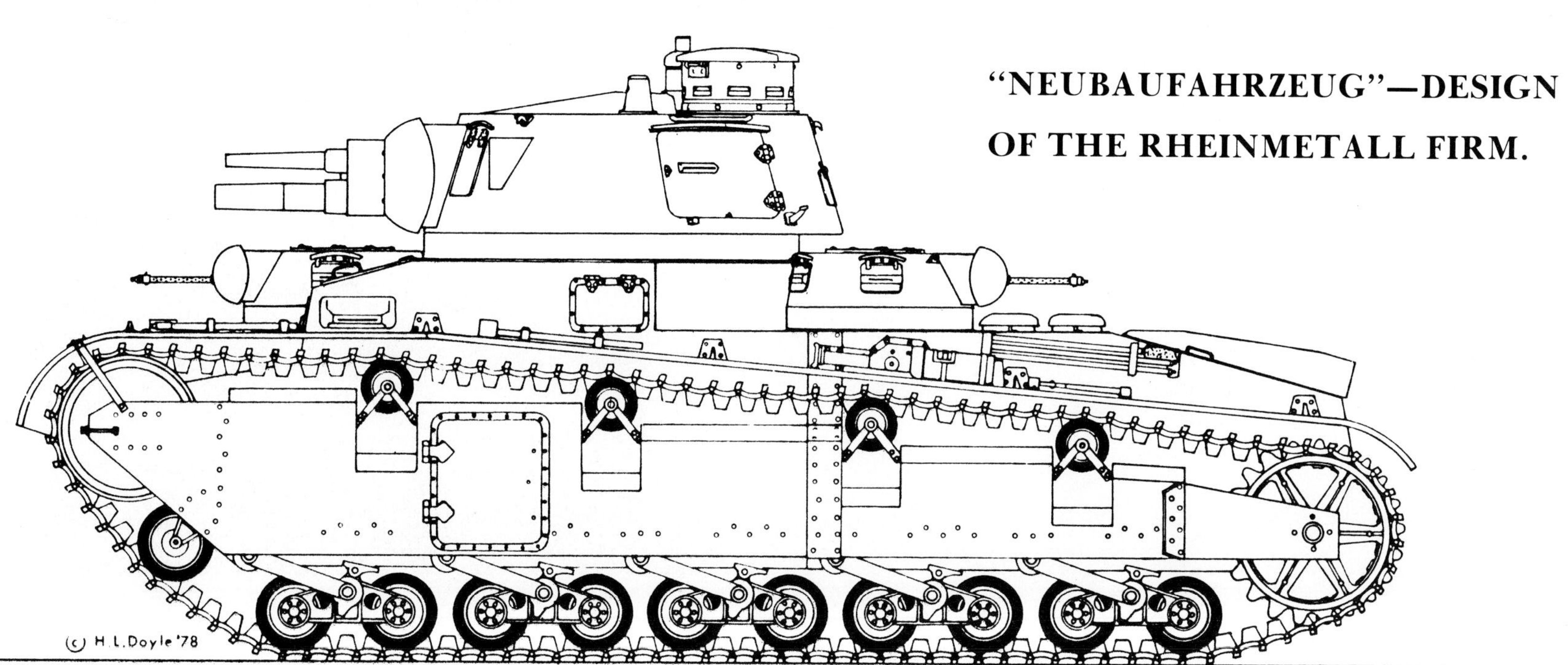

"NEUBAUFAHRZEUG"—DESIGN OF THE RHEINMETALL FIRM.

Technical Data

Neubaufahrzeug (Krupp)

Crew	6 men
Overall dimensions	6650 x 2900 x 2900
Armor plate	Front 20 mm, sides 13 mm, rear 13 mm
Armament	1 10.5 cm KwK, 1 3.7 cm KwK L/45, 2 MG
Ammunition	Quantity unknown
Gross weight	23 tons
Track width	380 mm
Ground pressure	0.69 kg/sq. cm
Motor	V-12 Maybach HL 108 TR
Horsepower	290 HP
Power to weight	12.6 HP
Fuel capacity	Unknown
Fuel consumption	Unknown
Top speed	30 kph
Range (on road)	120 km
Wading ability	Unknown
Ground clearance	57 cm
Crossing ability	Unknown
Climbing ability	Unknown

A "Neubaufahrzeug" (also called "Early Battle Tank V") with the turret of the Krupp firm—here at the International Automobile Exposition (Berlin, Spring 1939).

A "Neubaufahrzeug" of the Rheinmetall firm with guns mounted one above the other. Since its heaviest weapon was the 10.5 cm caliber gun, it is also called "Early Battle Tank VI".

BATTLE TANK I A

The Battle Tank I —Sd. Kfz. 101

By the Treaty of Versailles, Germany was banned from building tanks. Yet a camouflaged development—done partly in the Soviet Union—of light, medium and heavy tank types was carried on even in the Twenties. In this nomenclature it was the caliber, not the weight, that was decisive.

For reasons of technology and production, though, after the rejection of the Treaty of Versailles and proclamation of military increases in 1935, at first only Panzer I—and soon Panzer II as well—could be delivered to the troops in large quantities. For reasons of propaganda, and to inspire fear, this tank, which had been planned only as a training vehicle and not as a battle tank, was displayed at every opportunity. From 1935 on there was not a parade, Reichspartei event, harvest festival or large-scale drill without its participation. Since it was used as a driving-school vehicle without its superstructure, all prewar armored corpsmen were trained on this Panzer I chassis.

There were two main types of it: The Panzer IA with a Krupp motor and IB with a Maybach motor. They differed in their performance (60 as opposed to 100 HP), the number of road wheels and return rollers (four and three as opposed to five and four), and the raised leading wheel of IB. There were also command cars of both types, which had a fixed superstructure with just one machine gun in a round hatch instead of the rotating turret.

In all, the following numbers were built:

—IA 477
—IB 2000
—Command Car 200

They were integrated into the light companies of the armored regiments of the first five armored divisions and the independent armored units. The command cars were used as company chiefs' and commanders' cars in the staffs of the regiments and units.

When the war broke out too soon, they were—contrary to planning—also made operational as battle tanks. For instance, with the "Condor Legion" in Spain in 1936 and 1937, in Poland in 1939, Norway and France in 1940, and to some extent also in the Balkans and Russia in 1941. By 1942, though, there were no more Panzer I tanks in the armored battle units.

As early as 1939 efforts were made to replace them with better battle tanks as soon as possible. Thus the captured Czech 35 (t) and, as of 1940, the 38 (t) tanks, as well as ever-greater numbers of the German Battle Tank III, took their places. The Panzer I tanks recalled from the front now served only—sometimes until the war's end—as driving-school vehicles, training tanks, towing tractors, ammunition and weapons carriers, command cars for other types of units, and occasionally still as battle tanks in fighting against partisans in the hinterlands. The most notable types of these uses can be listed:

—Ammunition carriers
—Engineer Tank I
—4.7 cm antitank (t) gun carriers
—15 cm s.IG 33 gun carriers

Battle Tank I, Type A, in peacetime training.

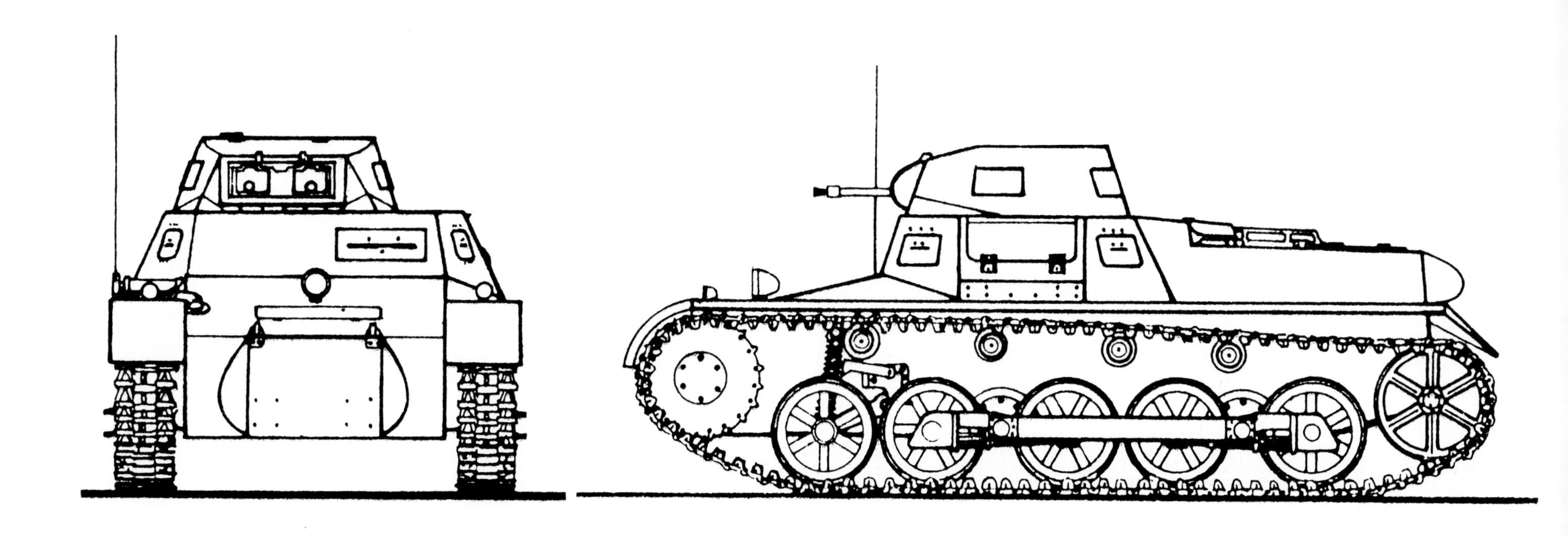
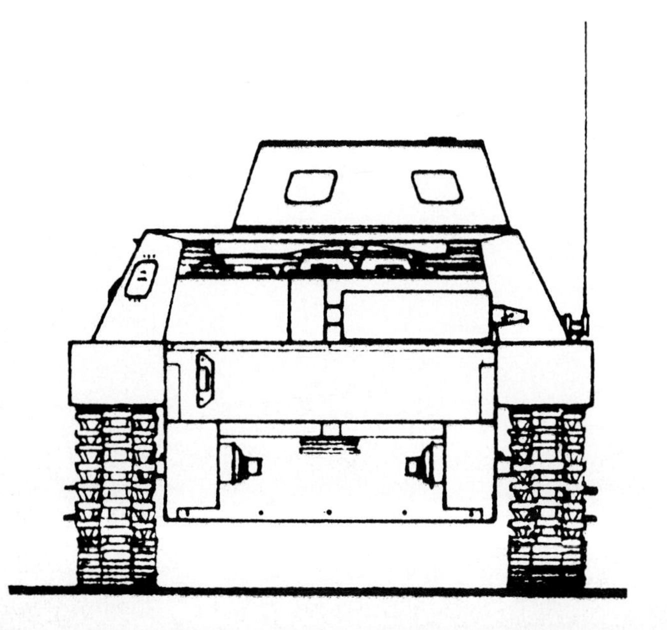
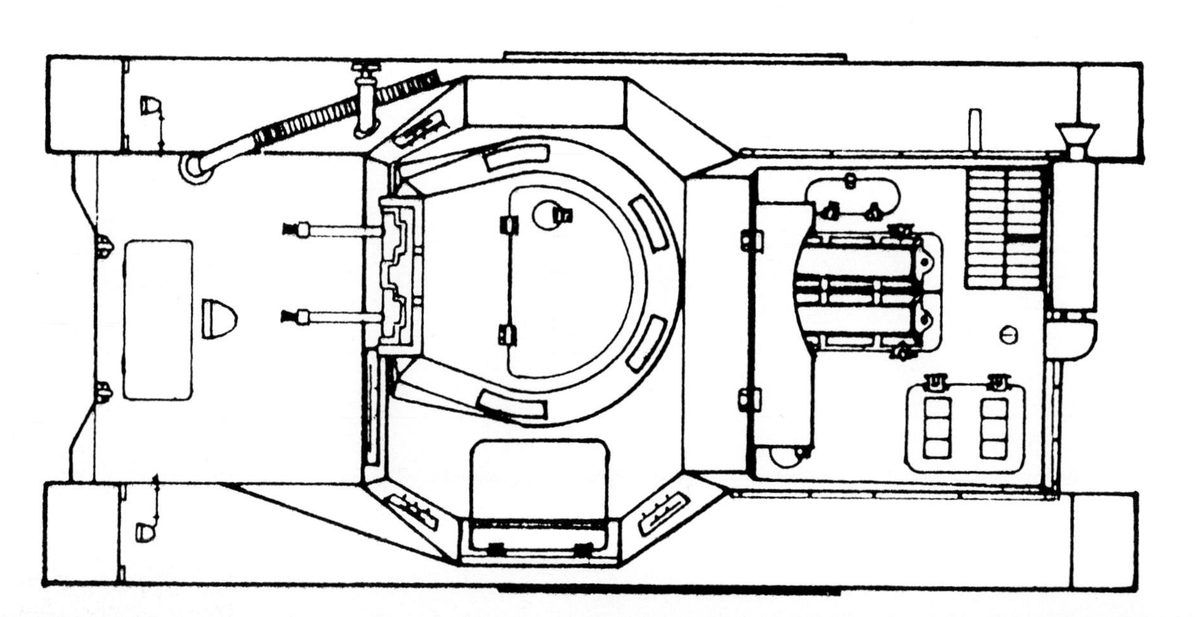

Technical Data

Battle Tank I, Type B

Crew	2 men
Overall dimensions	4420 x 2060 x 1960
Armor plate	Front 13 mm, sides 4 mm, rear 7 mm
Armament	2 machine guns
Ammunition	1525 machine gun rounds
Gross weight	5.9 tons
Track width	280 mm
Ground pressure	0.42 kg/sq. cm
Motor	Maybach NL 38 TR 6-cylinder
Horsepower	100 HP
Power to weight	17 HP
Fuel capacity	146 liters
Fuel consumption	80 liters/100 km (road)
Top speed	40 kph
Range (on road)	180 km
Wading ability	58 cm
Ground clearance	29 cm
Crossing ability	140 cm
Climbing ability	36 cm

Battle Tank I B units in a parade. This version is recognizable by its five instead of four road wheels, four instead of three return rollers, and the lack of exhausts on the track covers.

A Battle Tank I B command car.

The Battle Tank II—Sd. Kfz. 121

Barely a year—in 1935—after the first public appearance of the Panzer I, the Panzer II tanks, built by MAN, reached the troops. They were armed with a 2 cm machine cannon (BMK) and a machine gun.

Planned from the start as armored reconnaissance vehicles, they were first sent to the reconnaissance columns of the armored regiments and units. These columns consisted of five to seven Panzer II. But they were also used by the light armored companies in place of the Panzer III, which was planned as the standard tank but at first produced only in small quantities; there they were generally used in the form of five-tank columns, and sometimes also as column command cars for Panzer I columns.

The following models of it existed:

—II Type a and b,
—II Type c,
—II Type A-C,
—II Type D and E,
—II Type F.

While Types IIa and IIb had small road wheels, all the others from IIc on had the familiar five large disc wheels. From Type II A on there was an angled bow in place of the earlier rounded one, from Type B on an additional commander's cupola on the turret, from C on strengthened armor plate, and from F on a conical lead wheel.

Types IIc, C and F were most numerous. The last Panzer II of Type F was built in 1942.

From 1938 on there was one other development by Daimler-Benz as a "fast battle vehicle", with 180 instead of 140 horsepower. It is externally recognizable by the large road wheels without return rollers. Intended originally for the light divisions, it was nevertheless soon found in other units as well. There were Types D, E, G and I of it, scarcely to be told apart from the outside, apart from the further development of the "Luchs" (Lynx).

Its production was halted in 1941. In all, there were only about 250 examples of it. The Panzer II saw service with the Condor Legion in the Spanish Civil War and on all World War II fronts until the war's end—in the end, only in the reconnaissance columns of the armored regiments and units. It proved to be a reliable vehicle, aside from the weakness of its armament and the somewhat too low-powered motor in the first years of the war. Later, though, its armament and caliber were too weak to be able to withstand battle use.

With the end of production and replacement by stronger battle tanks, it served other types of units as command vehicles, but above all as weapons carriers. The best-known versions are:

—"Marder II" (Marten) with 7.62 cm antitank gun (r),
—"Marder II" with 7.5 cm antitank gun 40,
—"Wespe" (Wasp) with 10.5 cm leFH,
—"Bison" with 15 cm sIG 33, and
—"Luchs" (Lynx) armored scout car.

A Battle Tank II, Type A.

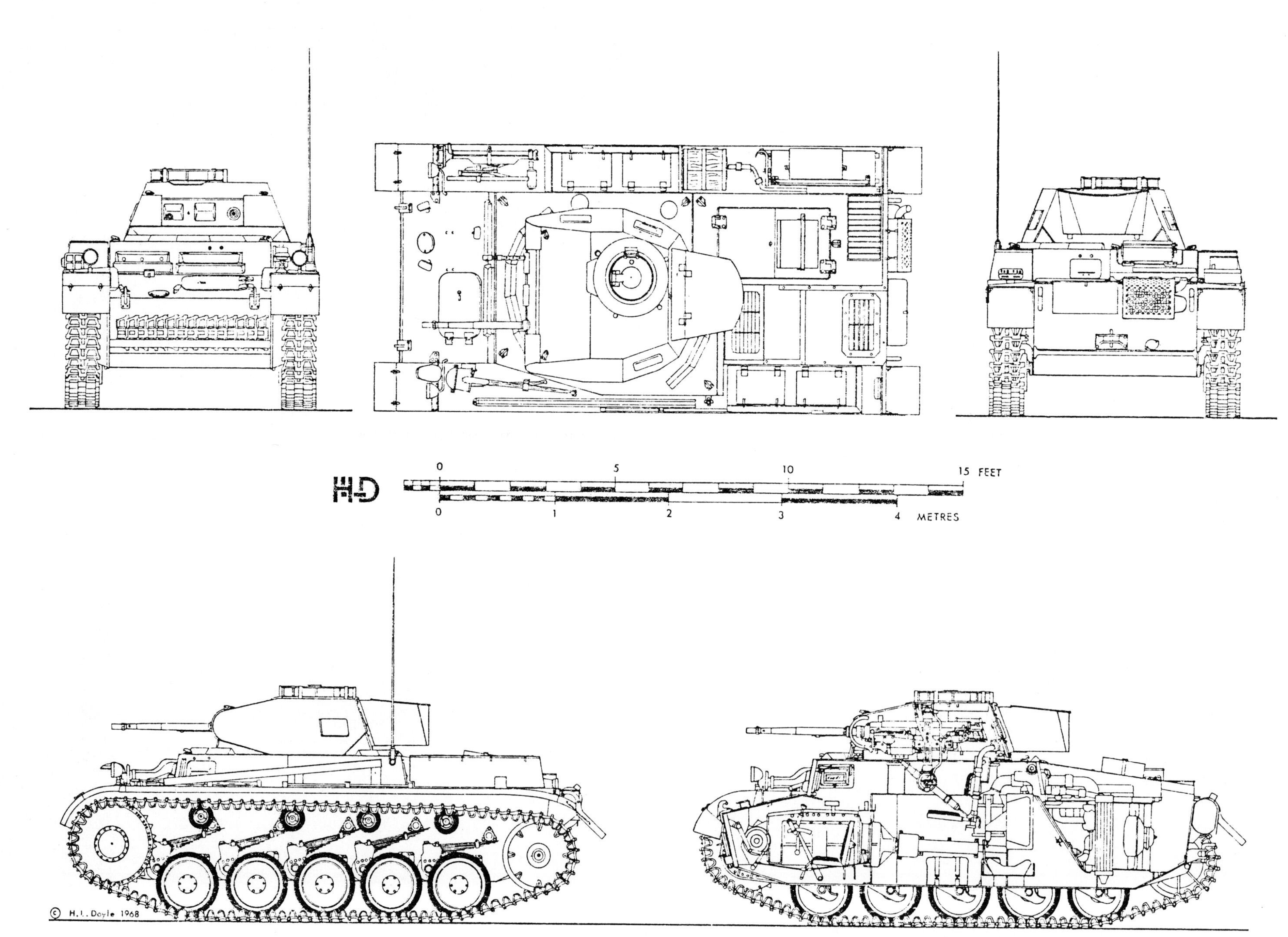

0
5
10
15 FEET
0
1
2
3
4 METRES
© H.L. Doyle 1968

Technical Data

Battle Tank II, Type F

Crew	3 men
Overall dimensions	4810 x 2280 x 2020
Armor plate	Front 35 mm, sides 14.5 mm, rear 20 mm
Armament	1 20 mm KwK 38 L/55. 1 machine gun
Ammunition	120 20-mm rounds, 2550 machine gun rounds
Gross weight	9.5 tons
Track width	300 mm
Ground pressure	0.66 kg/sq. cm
Motor	Maybach HL 62 TRM 6-cylinder
Horsepower	140 HP
Power to weight	14.7 HP
Fuel capacity	170 liters
Fuel consumption	110 liters/100 km (road)
Top speed	40 kph
Range (on road)	150 km
Wading ability	92 cm
Ground clearance	34 cm
Crossing ability	170 cm
Climbing ability	42 cm

A Battle Tank IIc—recognizable by its rounded bow. It was one of the most frequently built types.

Battle Tank II of the last (F) type—recognizable by its strengthened front end.

111

The Battle Tank 35 (t)

When the German Reich occupied Czechoslovakia in March of 1939, it took over the equipment of that country's armed forces. This included two types of tank (LTN 35 and TNHP) of advanced construction, both of them armed with a 3.7 cm cannon and two machine guns.

The somewhat older LTN 35 (35 ' its first production year) was designated 35 (t) (t for Tschechisch ' Czech), and turned over to the independent Armored Regiment 11 and to Armored Unit 65 of the 1st Light Division in the same year, as replacements for their Panzer I tanks. Since Armored Regiment 11 also joined the 1st Light Division at the beginning of the war, this division—as paradoxical as it sounds—became the strongest "Armored Division" of the Wehrmacht, even in the Polish campaign, thanks to the many 3.7 cm guns in its armored units—strongest in terms of firepower. It remained so in the French campaign too—though by now it was called the 6th Armored Division.

The 35 (t) was a robust tank with a technically good armament. The gears, brakes and steering, operated by compressed air, made the drivers' work much easier. Since the power train ran only a short distance from the motor to the rear wheels (the contemporary German tanks were driven by the front wheels), it was less likely to break down and had more space up front, so that after being taken over by the German armored troops, space was found in it for a radioman and his equipment (not provided in the Czech Army), despite the relatively small dimensions of the hull. Its one weakness worth mentioning was its riveting; when it took a direct hit on its armor plate, rivet heads were torn off and killed or wounded the crew.

This "Skoda 35 (t)"—as it was also known—remained in the armored units of the 6th Armored Division until the end of 1941. After that the remaining ones were given up, as this tank was not the equal of the T-34, KW I and II. While the majority of them were put to use by the allied Hungarians, Rumanians and Italians, and by German police units fighting partisans, others were rebuilt into Towing Tractor 35 (t) units (without turrets).

On the training grounds, February 1940.

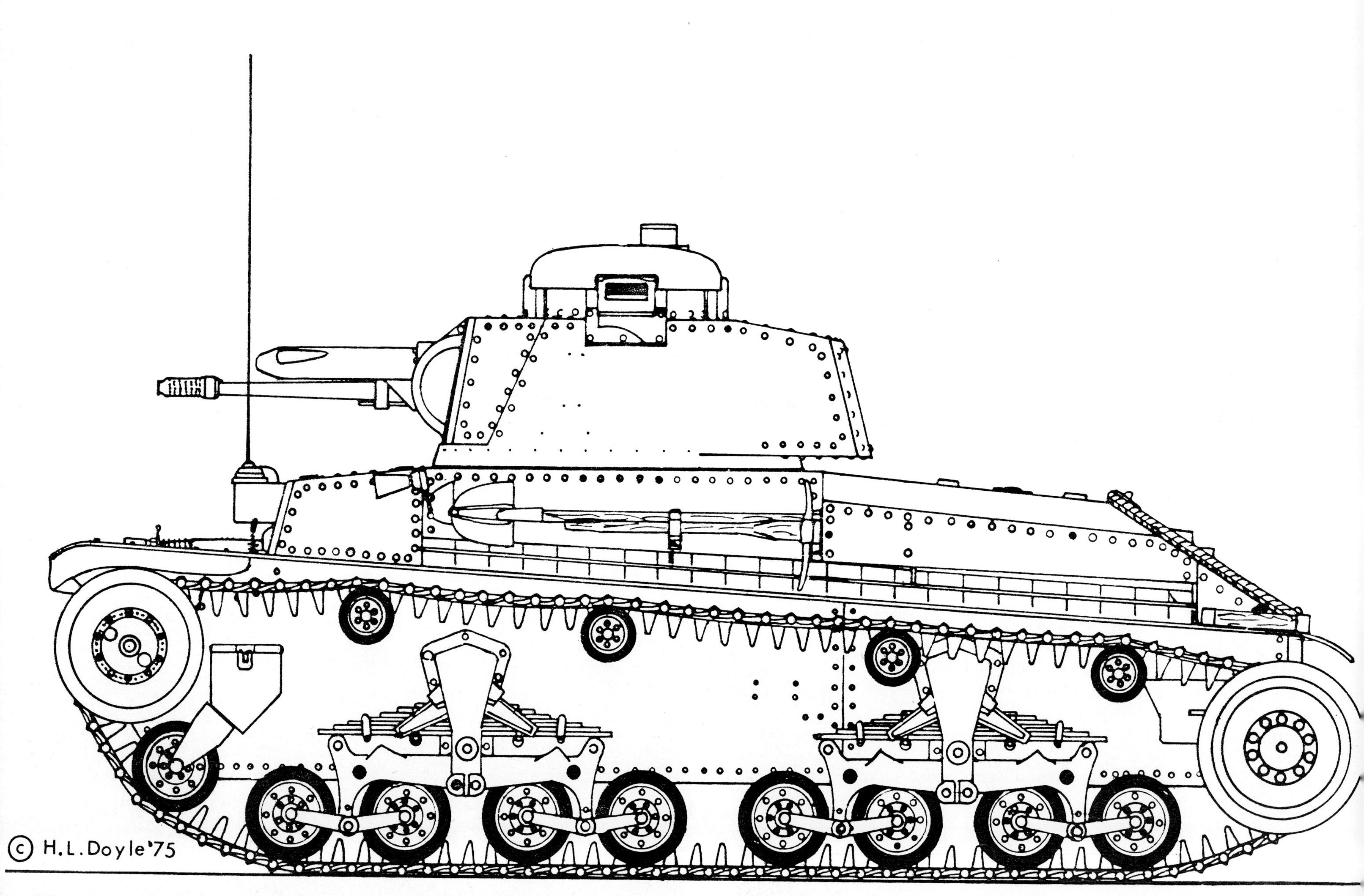
© H.L.Doyle'75

Technical Data

Battle Tank 35 (t)

Crew	4 men
Overall dimensions	4470 x 2130 x 2190
Armor plate	Front 25 mm, sides 16 mm, rear 12 mm
Armament	1 3,7 cm L/40, 2 machine guns
Ammunition	72 3.7 cm rounds, 1860 machine gun rounds
Gross weight	10.5 tons
Track width	320 mm
Ground pressure	0.52 kg/sq. cm
Motor	6-cylinder Skoda T 11
Horsepower	120 HP
Power to weight	11.4 HP
Fuel capacity	153 liters
Fuel consumption	80 liters/100 km (road)
Top speed	34 kph
Range (on road)	190 km
Wading ability	80 cm
Ground clearance	35 cm
Crossing ability	270 cm
Climbing ability	56 cm

Bow and stern of this battle tank, used almost exclusively in the 6th Armored Division (Armored Regiment 11 and Armored Unit 65).

Panzer 38 (t)

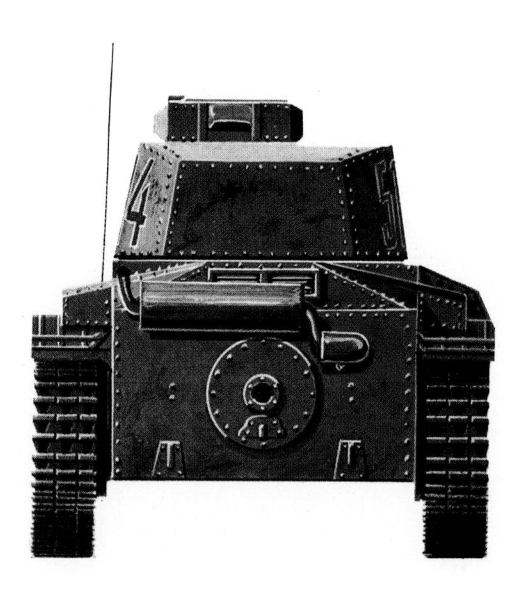

Battle Tank 38 (t)

Along with the Czechoslovakian LTN 35—in 1939—the Wehrmacht also took over the more modern TNHP, also known as LTL-H or LT 38. In the German armored forces it was known as the 38 (t).

It showed in even more improved form the modern design features of the 35 (t), but also had riveted armor plate. The armament was the same as the 35 (t), the motor somewhat more powerful (125 HP), later even reaching 150 HP. It was characterized by four large road wheels on each side.

Since it was still on its way to the Czech armored troops and only available in small numbers when the German takeover occurred, whole units could be equipped with it only after the Polish campaign. These were the 7th and 8th Armored Divisions. In them it replaced the German Panzer I and—when already on hand—the Panzer III.

Because of its good qualities, it had already been ordered by Sweden, Switzerland and Peru before Czechoslovakia was occupied; other countries showed great interest, but of course no deliveries were made after the German occupation. Its production was increased immediately, as Germany's allies were also supplied with it. So it was that about 800 38 (t) tanks began the campaign against Russia in 1941—almost one quarter of all German battle tanks. But its time ran out when the T-34, KW I and II appeared. Thus its production as a battle tank was halted in 1942.

In all, about 1500 of them were manufactured. Other than strengthening the armor plate and the aforementioned horsepower increase, there were only minor changes, scarcely visible from outside. This was a sign of its good quality from the beginning. Its proven chassis was still built in great numbers until 1944 and served the Wehrmacht to the end of the war as a weapons carrier of all kinds. These versions can be listed:

—Armored Ammunition Carrier 38 (t),
—Armored Command Vehicle 38 (t),
—"Bison" with 15 cm sIG 33, Types M and H,
—"Marder III" with 7.62 cm Pak (r) antitank gun,
—"Marder III" with 7.5 cm Pak 40 antitank gun, Types M and H.

Its chassis, in somewhat wider form, became the basis of the "Hetzer" armored pursuit vehicle. It was one of the most successful pursuit tanks of the last war, was produced in Czechoslovakia in scarcely changed form until the Fifties, and served there and in other countries (Switzerland and Sweden) until the Sixties. Thus the 38 (t) is probably the longest-lived tank in the world.

Up to eight series can be differentiated. Aside from a more powerful motor, there were mainly strengthenings of the armor.

Battle Tank 38 (t), Last Version

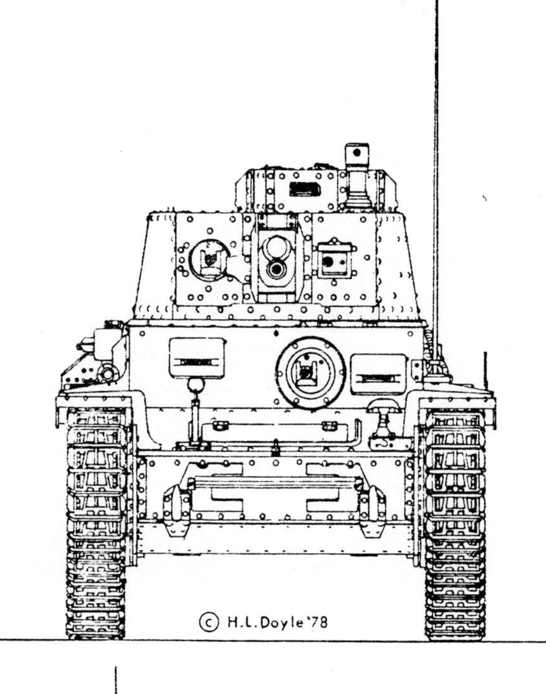

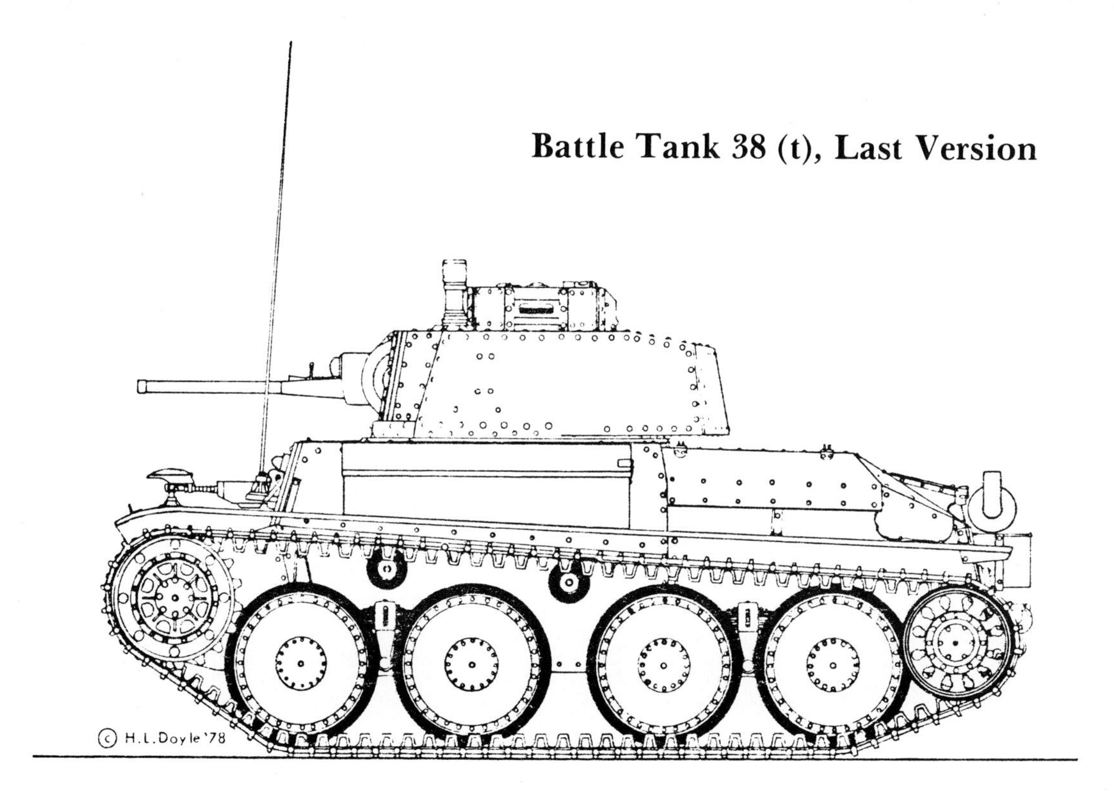

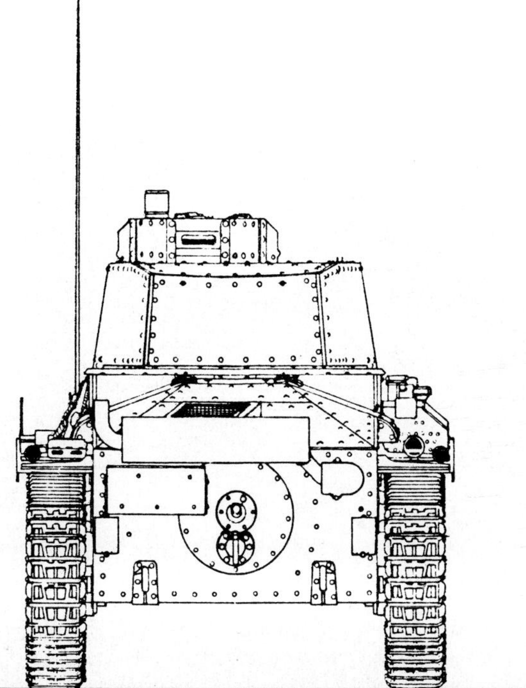

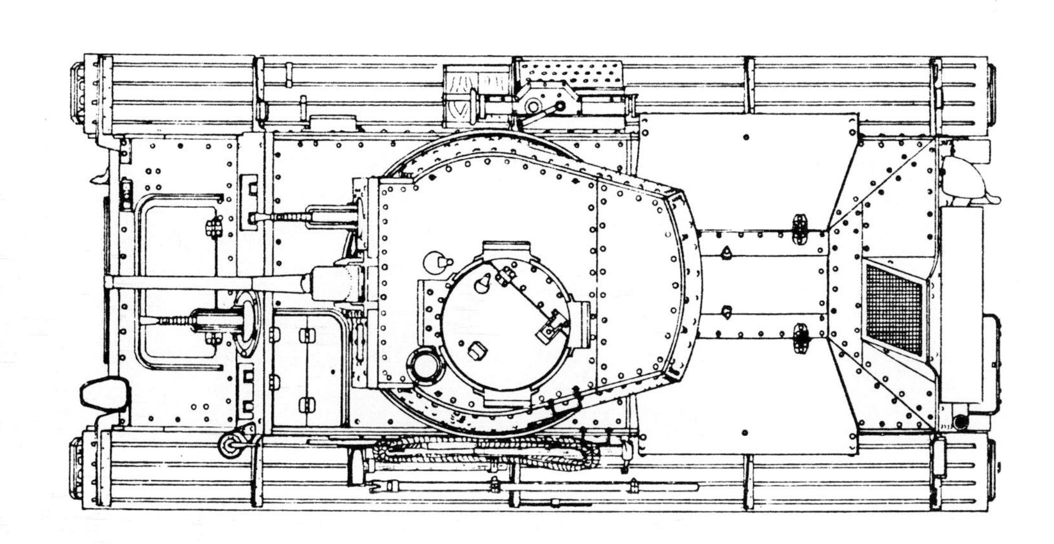

Technical Data

Battle Tank 38 (t) Series E

Crew	4 men
Overall dimensions	4880 x 2130 x 2130
Armor plate	Front 25 mm, sides 15 mm, rear 12 mm
Armament	1 Skoda 3.7 cm L 40, 2 machine guns
Ammunition	42 3.7 cm rounds, 2400 machine gun rounds
Gross weight	10.5 tons
Track width	293 mm
Ground pressure	0.57 kg/sq. cm
Motor	6-cylinder Praga EPA
Horsepower	150 HP
Power to weight	14.3 HP
Fuel capacity	218 liters
Fuel consumption	90 liters/100 km (road)
Top speed	42 kph
Range (on road)	250 km
Wading ability	90 cm
Ground clearance	40 cm
Crossing ability	260 cm
Climbing ability	53 cm

Above is a Battle Tank 38 (t) of Series B (or C?), below one of Series S. The strengthening of the latter's front end is obvious.

BATTLE TANK III—TYPE B

The Battle Tank III (short)—Sd. Kfz. 141

In 1934 the army requested a medium battle tank that was planned as a support or command tank for the Panzer I and II. It was to be equipped with an armor-penetrating cannon as well as additional radio equipment. The first Panzer III could be delivered to the troops as early as 1936. But instead of the ordered 5 cm KwK, they carried only—since it was already available—a 3.7 cm KwK. Along with that were three (later only two) machine guns. The crew consisted of five soldiers.

In all armored units, it was supplied mainly to the medium (4th to 8th) companies, and more rarely to the light companies as a column-leading tank. It was soon found in all staffs as a command vehicle, with additional radio equipment and a wooden dummy gun—in order to make room for the map board in the turret.

No other German army tank was the subject of as many experiments as the Panzer III. In all, there were twelve versions of it, of which nine (A through J) had the short 3.7 cm KwK and later the 5 cm type. There were also command vehicle versions of Types D and E. The types varied as follows:
Type A (1936): five large road wheels on each side.
Type B (1937): eight small road wheels on each side.
Type C (1937): like B but with a new leaf spring arrangement.
Type D (1938): angled leaf springs, new commander's cupola.
Type E (1939): six road wheels on each side.
Type F (1940): minor changes from Type E. This and Type E were built in greatest numbers with the 3.7 cm gun.
Type G (1941): turret lengthened to the rear and usually armed with a (short) 5 cm KwK. During reequipping as of 1941, many Type E and F Battle Tank III were also armed with a 5 cm KwK.
Type H (1941): changed leading and drive wheels (spoked type). It was the most often built Panzer III with a short gun.
Type J (1942): Basic armor increased from 30 to 50 cm.

With the withdrawal of the Panzer I, it became as of 1940 the main battle tank of the light companies and thus the basic tank of the German armored troops. In all, about 3000 (short!) of them were built. It saw service on all fronts up to 1942—including North Africa. It proved to be a very well-developed battle tank, though in comparison to its weight and dimensions it was short on firepower, even with the 5 cm (short) KwK. Its days therefore ended with the appearance of the Russian T-34. It was then equipped with a long (L/60) 5 cm KwK and will be described in the next chapter.

In its later versions (from E on) there were many mixed types, on account of re-equipment and repair. But its chassis already served quite early for special versions such as:
—Assault Gun III,
—"Bison" with the sIG 33,
—Armored Ammunition Carrier III.

Battle Tank III, Type F, with the readily visible strengthened armor.

TYPE A

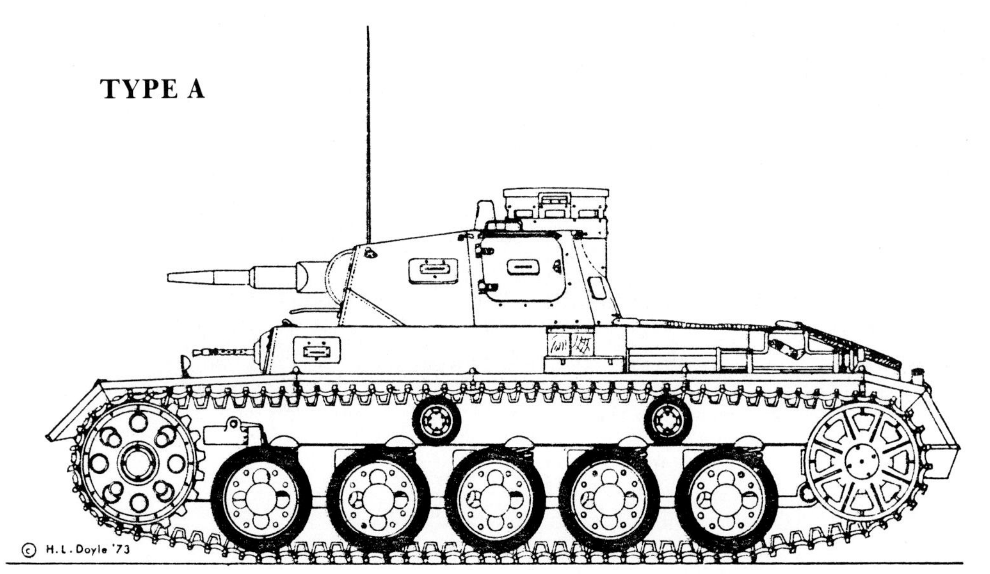

TYPE D

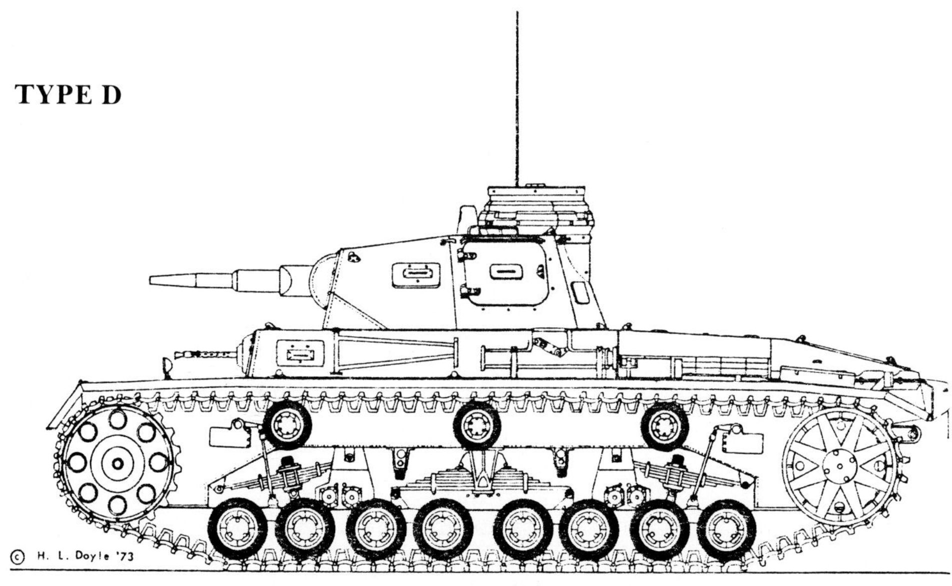

TYPE F

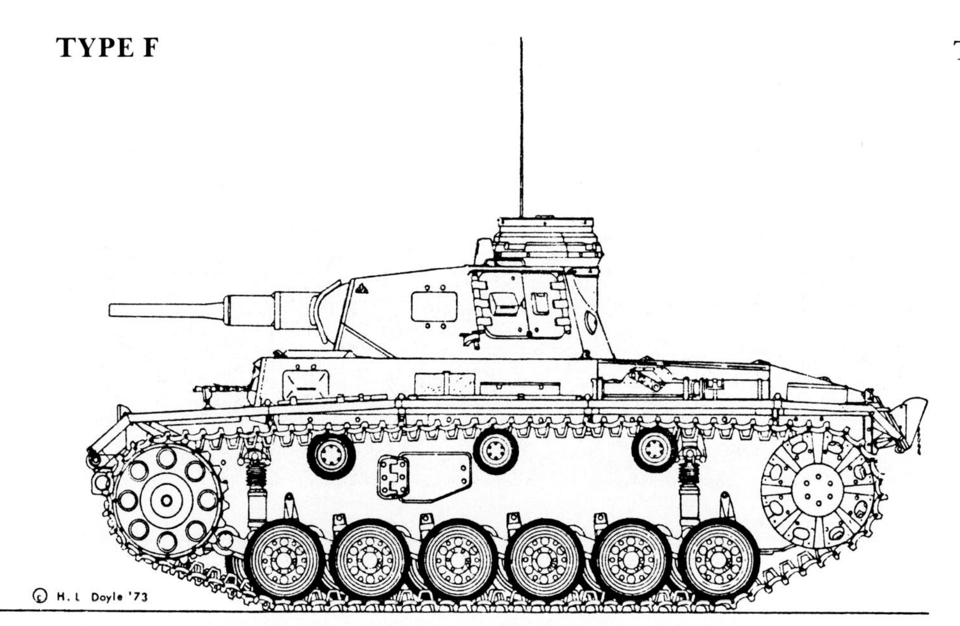

TYPE J

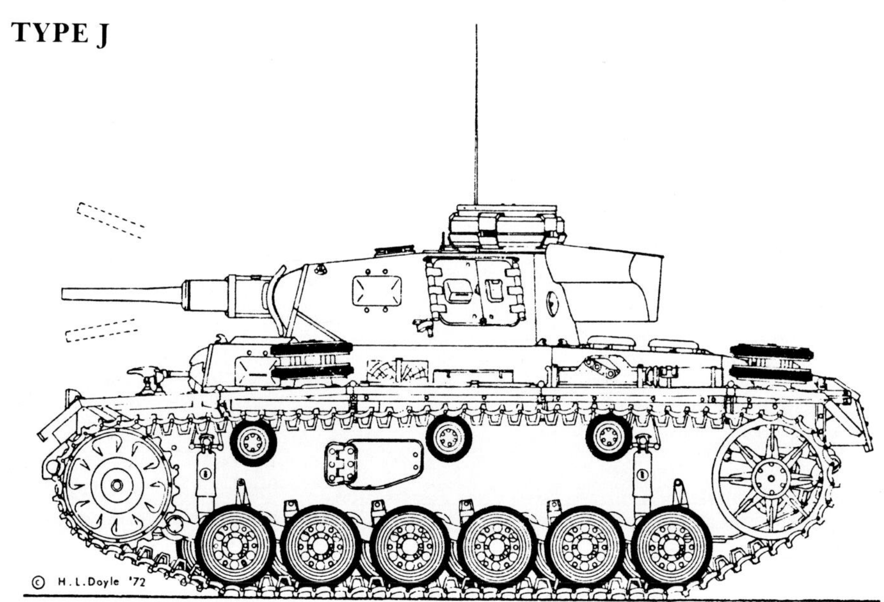

Technical Data

Battle Tank III Type H

Crew	5 men
Overall dimensions	5520 x 2950 x 2500
Armor plate	30 mm all around
Armament	1 5 cm KwK L/42, 2 machine guns
Ammunition	99 5 cm rounds, 2700 machine gun rounds
Gross weight	21.6 tons
Track width	400 mm
Ground pressure	0.99 kg/sq. cm
Motor	Maybach HL 108 TR 12-cylinder
Horsepower	265 HP
Power to weight	12.2 HP
Fuel capacity	320 liters
Fuel consumption	220 liters/100 km (road)
Top speed	40 kph
Range (on road)	140 km
Wading ability	80 cm
Ground clearance	38 cm
Crossing ability	259 cm
Climbing ability	60 cm

A Type E with the 3.7 cm KwK.

This Type H with the 5 cm KwK already has the changed shield and a strengthening plate on the front end. It is the most often built version of the Battle Tank III (short).

BATTLE TANK III
TYPE L

The Battle Tank III (long)—Sd. Kfz. 141/1

As one answer to the Russian T-34, the Battle Tank III, Type J was armed, from 1942 on, with a longer (L/60) 5 cm KwK. Of this type, now called Battle Tank III (long), the following versions were built before the end of production in 1943:
Type J (1942): New sight visors, new gun shield, storage rack on the turret.
Type K (1942): Existed only as a command vehicle.
Type L (1942): Additional 20 mm armor via offset watertight armor at the front end and in front of the shield. Fighting weight now 23 tons.
Type M (1942/43): Modified exhaust system and armor skirts around the turret and on the sides of the hull.

As with the "short" type, there were also mixed versions resulting from reequipping and repair. In all, more than 2000 of these strengthened Panzer III were built. Beginning as early as the beginning of 1943 they were replaced more and more by the Battle Tank IV (long), since even the 5 cm (L/60) KwK was not equal to the enemy tanks with their ever-increasing firepower. It was used mainly in Russia, but some saw service in North Africa.

The versions noted above served as the basis for the following special versions:
—Type N (1943): instead of the 5 cm KwK, it carried a 7.5 cm (L/24) gun, which had become available when the Battle Tank IV (short) was modified to (long).
—Several Type M were rebuilt into Flamethrowing Tank III.
—Assault Gun 40.
—Assault Howitzer 42.
—Observation Tank III.
—Recovery Tank III.

Aside from Panzer I, this was the first mass tank of the German armored troops. Yet its production had to be halted when the enemy appeared on the battlefield with larger calibers, longer guns (greater shot power) and heavier armor. It should have been replaced even earlier, but in 1942 there were not yet enough Battle Tank IV available with the long 7.5 cm KwK. In this light, all variations from 1942 on were merely stopgaps. From the beginning too, it was always under-armed in comparison to the Panzer IV, its equal in weight and nearly so in dimensions. In hindsight, it would have been better if only the Panzer IV had been built from the start.

In all, barely 6000 Battle Tank III (short and long) were built, including command vehicles.

A Battle Tank III (long) of Type J, the first that gave the "Long" its typical appearance (storage case on the turret, long gun barrel, later side skirts).

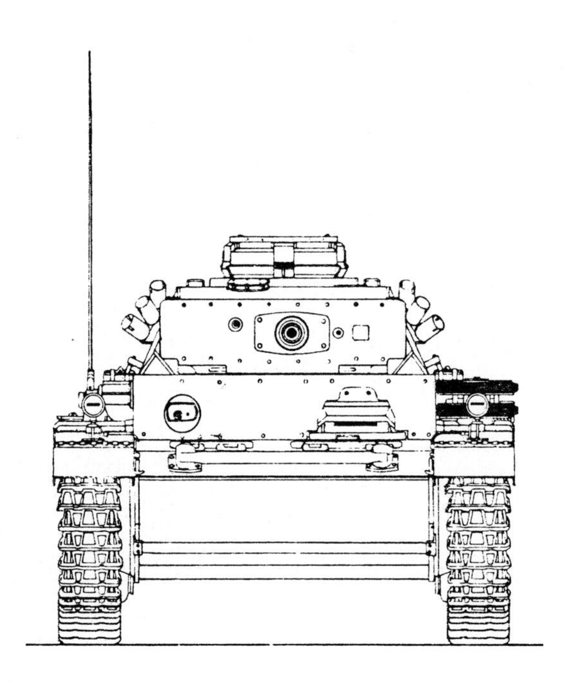

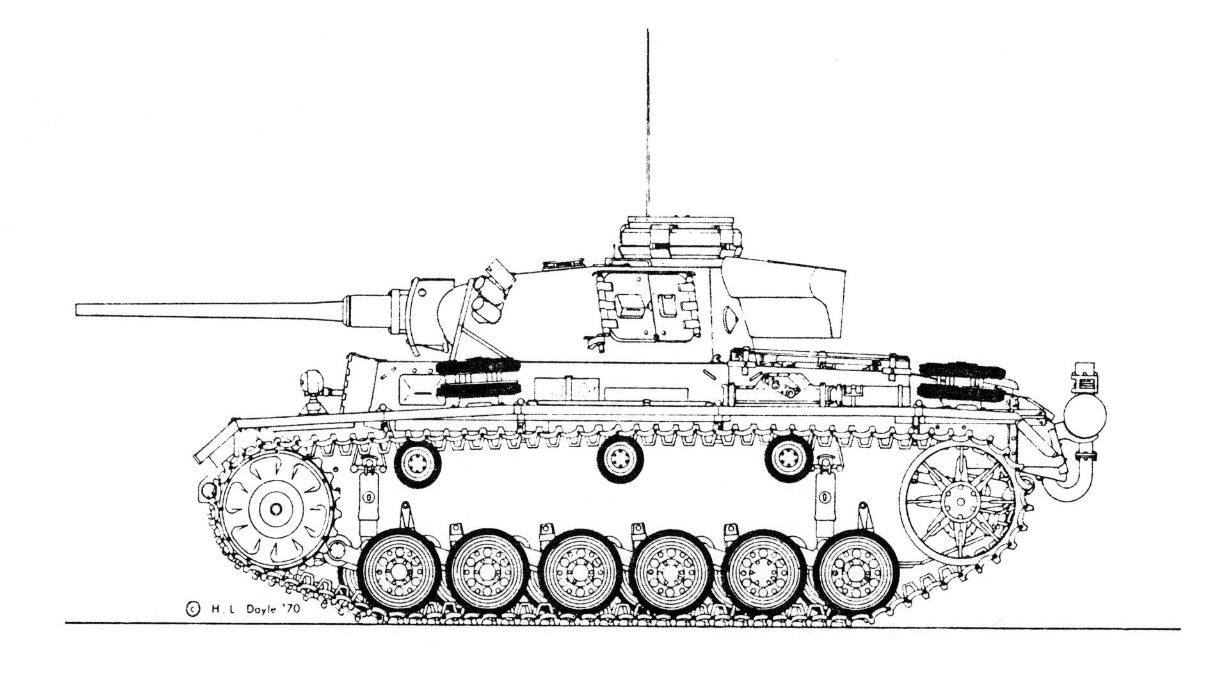

0 5 10 15 FEET

0 1 2 3 4 METRES

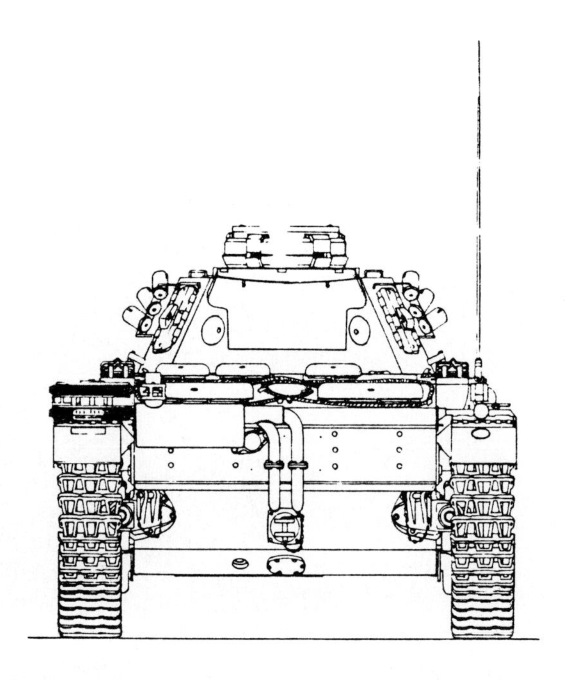

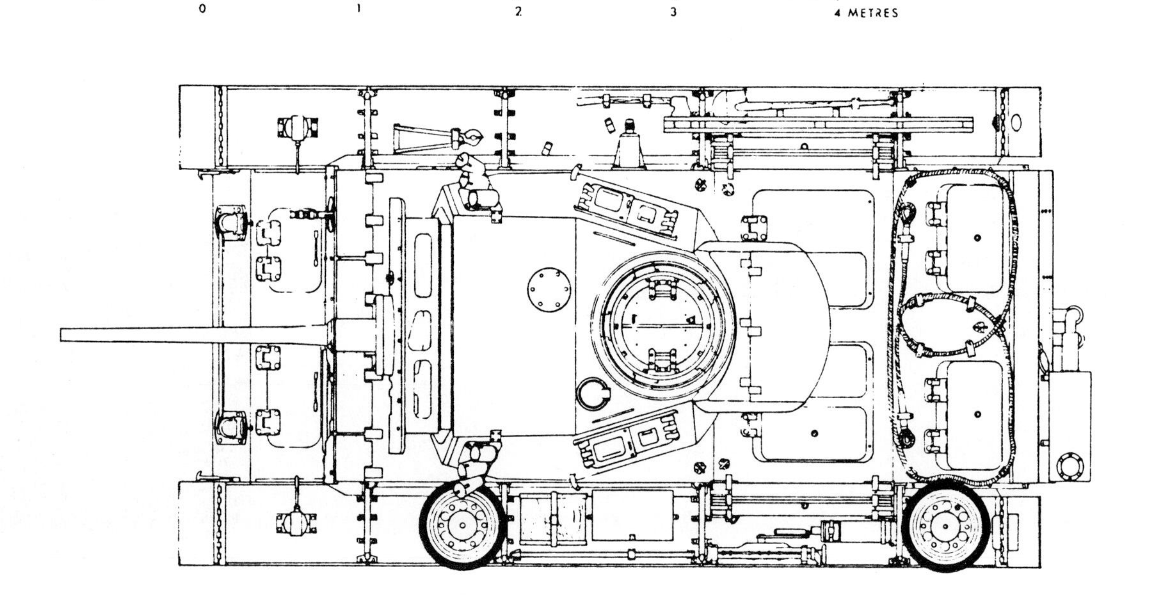

Technical Data

Battle Tank III, Type M

Crew	5 men
Overall dimensions	6412 x 2950 x 2500
Armor plate	Front 70 mm, sides 30 mm, rear 50 mm
Armament	1 5 cm KwK 39 L/60, 2 machine guns
Ammunition	92 5 cm rounds, 3750 machine gun rounds
Gross weight	22.3 tons
Track width	400 mm
Ground pressure	0.94 kg/sq. cm
Motor	Maybach HL 120 TRM 12-cylinder
Horsepower	265 HP
Power to weight	12.3 HP
Fuel capacity	320 liters
Fuel consumption	182 liters (road)
Top speed	40 kph
Range (on road)	155 km
Wading ability	80 cm
Ground clearance	41 cm
Crossing ability	259 cm
Climbing ability	60 cm

This Battle Tank III is also a Type J.

A Type L (or M?), recognizable by its armoring in front of the gun shield and on the front end.

The Battle Tank IV (short)—Sd. Kfz. 161

Planned since the early Thirties as an armored surveillance vehicle, and thus a parallel development to the subsequent "Newly Built Vehicle", the first 35 of this medium battle tank—now called Panzer IV—were delivered by Krupp in 1936. They were armed with a 7.5 cm KwK and two machine guns. The crew, like that of Panzer III, consisted of five men. Though it was somewhat wider and longer (with eight road wheels instead of six) and its turret was somewhat bigger, it still resembled the Panzer III greatly, especially as their components were interchangeable. Thus even before the war the question arose as to why two so similar tanks, differing almost exclusively in their armament, existed.

The first version was soon followed by Type B, and since it was built until the war's end, there were subsequent versions up to Type J. The main differences will be seen in the list below. As with Battle Tank III, at first all Battle Tank IV were armed with the short (L/24) 7.5 cm KwK—and called Panzer IV (short); they were:

Type A (1936).

Type B (1936): Somewhat higher hull, unangled front end, no machine gun for the radioman, new commander's cupola.

Type C (1937): Coaxial machine gun protected by armor, new motor.

Type D (1939): Bow machine gun and angled front end again, new gun shield. Additional front armor as of 1940.

Type E (1940): Additional side armor, new sight shield, new gun shield for the bow machine gun, reinforced commander's cupola.

Type F (1941): Unangled front end again, further armor strengthening, wider tracks.

With Type F the Battle Tank IV attained its final form. Subsequent variations consisted only of armor strengthening, the installation of a longer gun, and the armor skirts that became universal as of 1943. With the installation of the longer gun (see the next chapter), it was designated F2 and the remaining Type F tanks were called F1. During major servicing, though, these Panzer IV F1 types were also equipped with a longer cannon and renamed F2.

The Panzer IV (short) were used in the medium (sometimes erroneously called "heavy") companies. Every armored unit of them had one—always the 4th (or 8th in the second unit of a regiment). When the Panzer I was replaced by the Battle Tank III, also in service with these companies in the beginning, they had only Battle Tank IV as of 1941-42. They were divided into columns of three Panzer IV (short) each.

The Panzer IV (short) saw service in all campaigns—except Norway—including North Africa, from the beginning of the war to 1941. It proved itself well thanks to its well-developed chassis and motor. The cannon was effective enough against weak targets. Against enemy battle tanks, though, it was almost ineffective on account of its too-slow firing speed (' lack of penetrating power) and the thereby limited range of the shell (' lack of accuracy, especially against moving targets). Faced with the T-34, which could penetrate German armor at 1200 meters with its long cannon, the Battle Tank IV (short) soon met its end. As of November 1941, it was rearmed with a long 7.5 cm KwK.

In all, over 700 Panzer IV (short) were built. There were no special models of Types A through F.

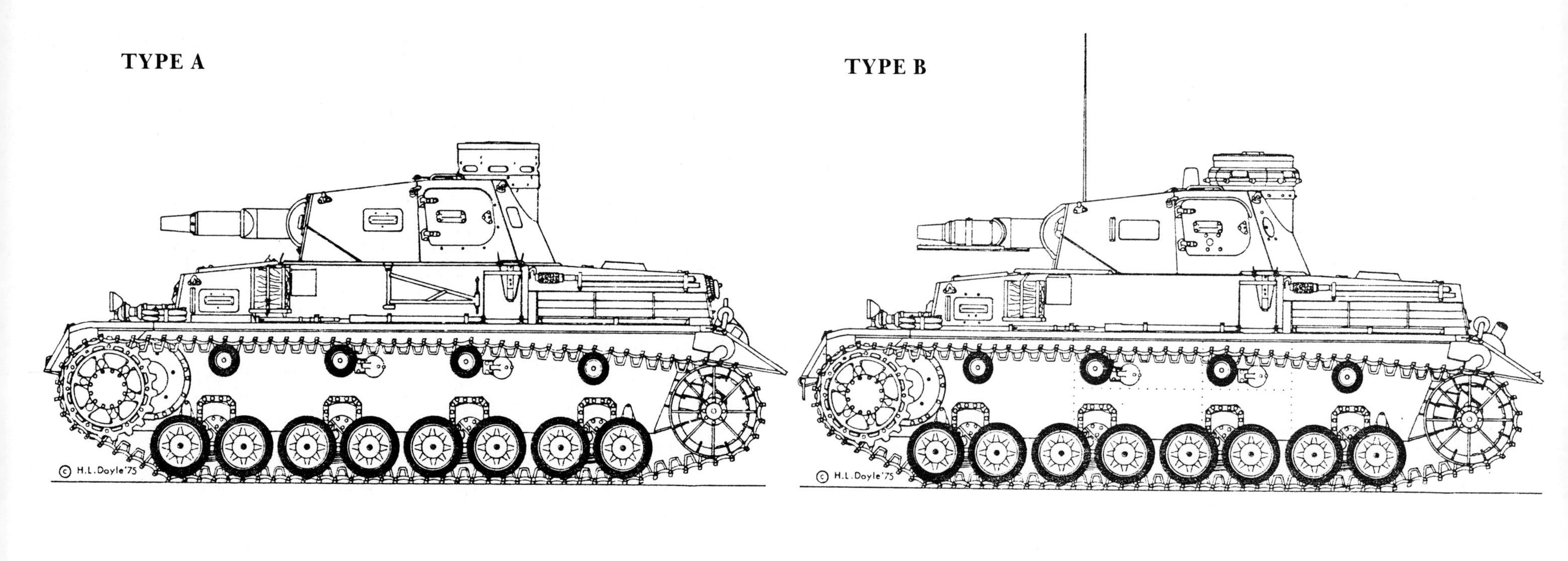
TYPE A
© H.L.Doyle'75
TYPE B
© H.L.Doyle'75

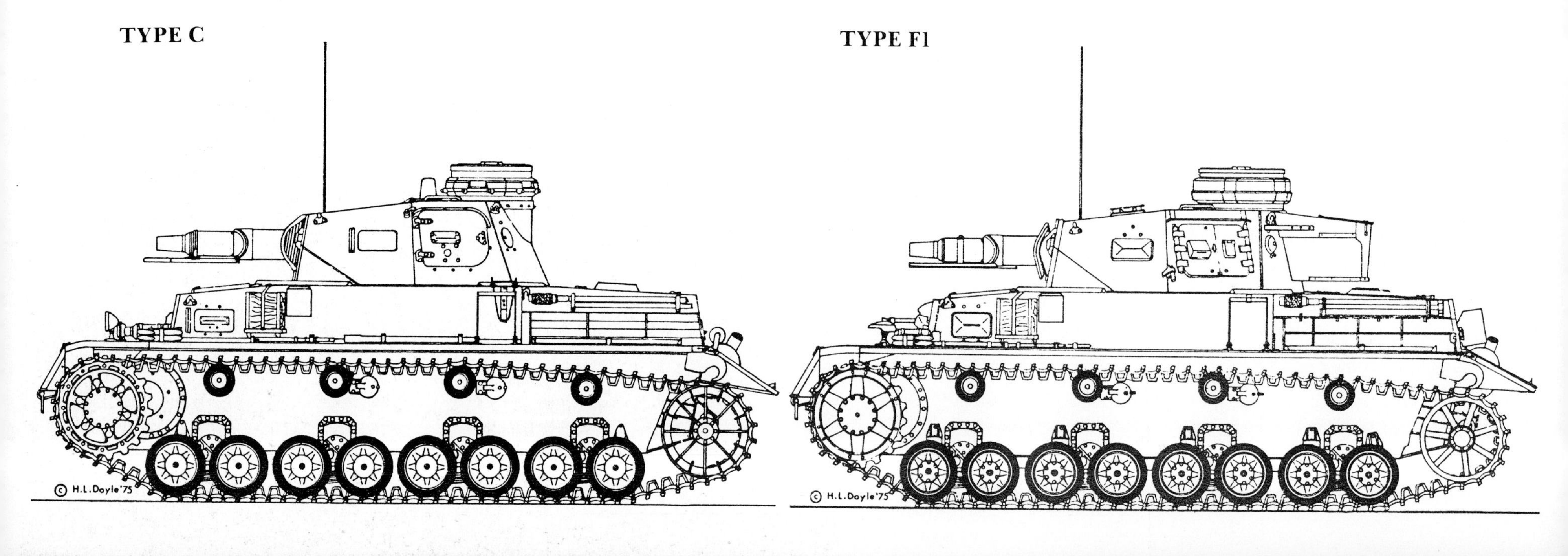
TYPE C
© H.L.Doyle'75
TYPE F1
© H.L.Doyle'75

Technical Data

Battle Tank IV Type E

Crew	5 men
Overall dimensions	5920 x 2860 x 2680
Armor plate	Front 30 mm, sides and rear 20 mm
Armament	1 7.5 cm KwK L/24, 2 machine guns
Ammunition	80 7.5 cm rounds, 2700 machine gun rounds
Gross weight	21 tons
Track width	380 mm
Ground pressure	0.79 kg/sq. cm
Motor	Maybach HL 120 TRM 12-cylinder
Horsepower	265 HP
Power to weight	12.6 HP
Fuel capacity	470 liters
Fuel consumption	200 liters/100 km (road)
Top speed	42 kph
Range (on road)	230 km
Wading ability	100 cm
Ground clearance	40 cm
Crossing ability	220 cm
Climbing ability	60 cm

Type A, recognizable by the angled front end, the thinly armored sight visors and the cylindrical turret.

Here, on the other hand, is the last short version: Battle Tank IV F1.

BATTLE TANK IV, TYPE F2 (A Type H is shown on the inside front cover.)

The Battle Tank IV (long)—Sd. Kfz. 161

With the installation of the long 7.5 cm L/43 KwK, the Panzer IV took on its modern appearance as of 1942. It was now at least the equal of the T-34. As of 1943 it was given an even longer (L/48) KwK, and was used more and more to replace the Battle Tank III, which had become unsatisfactory on account of its too-small caliber. Thus it became the mass tank of the war's last years. Since the Battle Tank V (Panther) that was intended toreplace it suffered from teething troubles and could not be produced in sufficient numbers because of the increasing bombing attacks, production of the Battle Tank IV continued to the war's end.

There were just four versions of it. Their basic differences are as follows:
Type F2 (1942): The long gun, plus new drive and leading wheels.
Type G (1942): Fewer sight visors on the turret, new muzzle brakes.
Type H (1943): Strengthened armor on the bow, longer (L/48) gun, antenna moved to the rear, new drive wheels, armor skirts.
Type J (1944): Elimination of the auxiliary motor for rotating the turret; the last series had cast leading wheels and steel return rollers.

There were numerous mixed versions of these types too, resulting from major repairs and reequipping. In all, about 8700 Battle Tank IV (long) were built.

From 1943 on it was in service with all armored companies that were not equipped with Panthers, Tigers or—there were some—assault guns. A special version of it, for use as a command car, was never built. When it was used as such, it was equipped with additional radio equipment—externally recognizable by its two or three antennae. It saw service on all front from 1942 on.

Basically, the Battle Tank IV—whether short or long—retained its form from 1936 to 1945, and since it saw service after the war in Spain, Turkey and Syria (there until 1967!), it also ranks among the longest-lived battle tanks in the world. German production added up to over 9500 short and long Panzer IV.

There were many special versions of its later types. Its chassis was also used particularly as a weapons carrier for the armored pursuit, artillery and anti-aircraft troops. The most important variants were:
—Assault Gun IV,
—Pursuit Tank IV,
—Assault Tank IV "Brummbär" (Growling Bear),
—"Hummel" (Bumblebee, 15 cm armored howitzer),
—"Nashorn" (Rhinoceros, 8.8 cm antitank gun),
—"Möbelwagen" (Moving Van, four 2 cm AA guns),
—"Wirbelwind" (Whirlwind, four 2 cm AA guns),
—"Ostwind" (East Wind, 3.7 cm AA gun).

The last type (J) of Battle Tank IV (long) made.

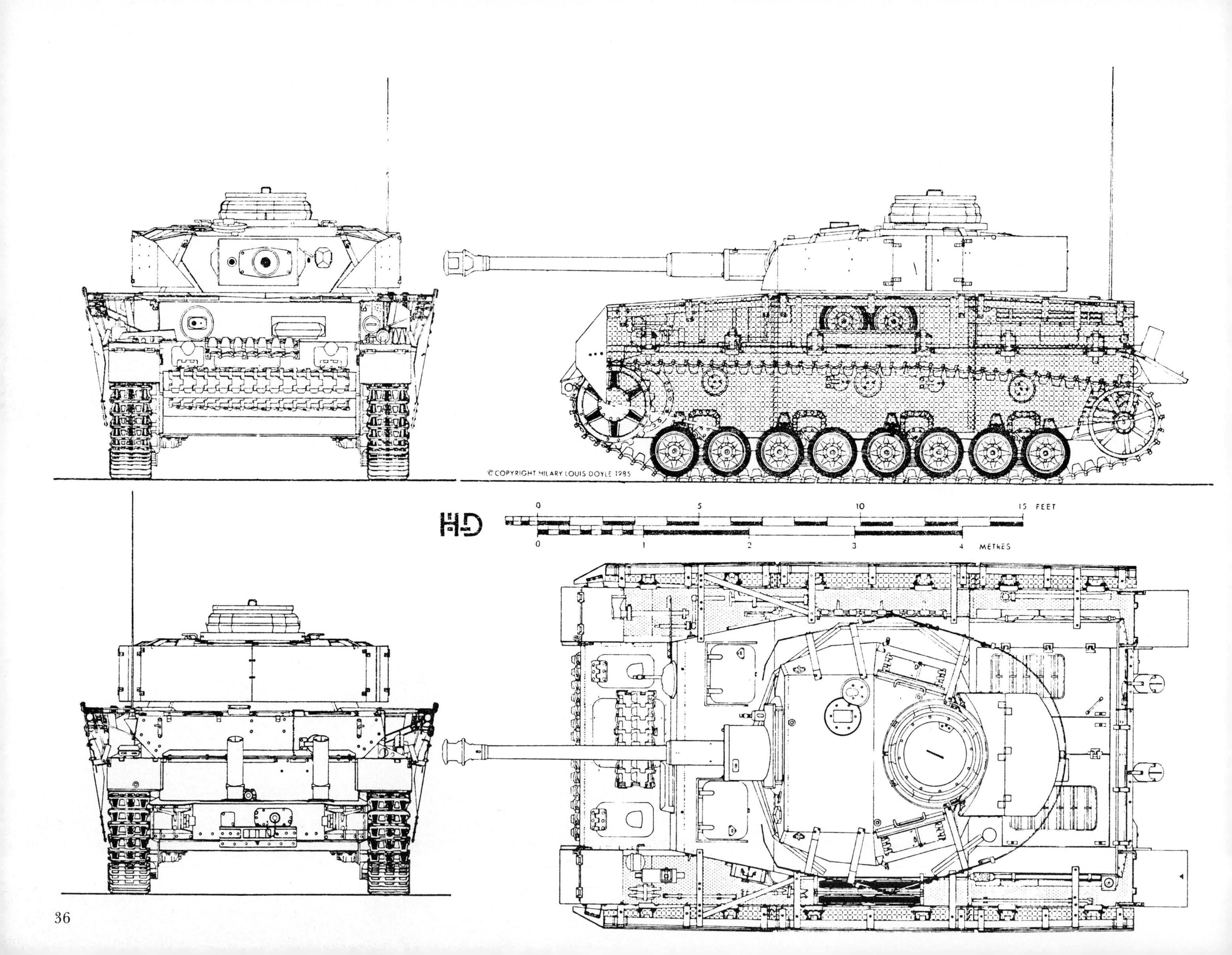
© COPYRIGHT HILARY LOUIS DOYLE 1985
HD
0
5
10
15 FEET
0
1
2
3
4 METRES

Technical Data

Battle Tank IV Type H

Crew	5 men
Overall dimensions	7015 x 2880 x 2680
Armor plate	Front 80 mm, sides 30 mm, rear 20 mm
Armament	1 7.5 cm KwK L/48, 2 machine guns
Ammunition	87 7.5 cm rounds, 3150 machine gun rounds
Gross weight	25 tons
Track width	400 mm
Ground pressure	0.89 kg/sq. cm
Motor	Maybach HL 120 TRM
Horsepower	265 HP
Power to weight	10.6 HP
Fuel capacity	470 liters
Fuel consumption	250 liters/100 km (road)
Top speed	38 kph
Range (on road)	180 km
Wading ability	120 cm
Ground clearance	40 cm
Crossing ability	235 cm
Climbing ability	60 cm

The Type F2 shows the somewhat shorter 7.5 cm KwK (long) with caliber length L/43. It can also be recognized by the spherical muzzle brake.

A Type H with skirts around the turret and on the hull, the longer (L/48) gun and the more cylindrical muzzle brake.

BATTLE TANK V, TYPE A

The Battle Tank V—Panther—Sd. Kfz. 171

With the surprising appearance of the new Russian battle tanks with their long 7.62 cm guns (which resulted in greater accuracy and penetrating power for the shells), their heavier armor, wider tracks and, especially in the T-34, almost ideal shot-deflecting designs, the end was in sight for the then existing German tank types. The call for at least equal German battle tanks could not be ignored. To save time, General Guderian even urged the copying of the T-34—though with better optics and radio equipment. But it was decided to develop a fully new type to replace the existing battle tanks and thus become the German mass tank.

But that took time for development, and so it was the summer of 1943 before the first—they were designated Battle Tank V and called "Panther"—appeared on the battlefield (Operation "Citadel", July 1943). As was to be expected, they showed many teething problems, and what with the shortage of time, training their crews and servicing staffs had its faults. As a result, more Panthers fell victim to engine fires and other reasons than to enemy action in this first operation. But the problems were soon solved, so that the Panther, going through two further stages, became one of the most successful battle tanks by the war's end and is still regarded today as the best of World War II.

The three types differ as follows:
Type D (1942): Cylindrical commander's cupola, no bow machine gun.
Type A (1043): New cast, shot-deflecting commander's cupola, bow machine gun in a round shield.
Type G (1944): Driver's sight visor in the angled bow armor eliminated; the upper periscope remains and now turns.

All three types also differed in having varying arrangements of the exhaust system. There were particularly command vehicles of all three types, recognizable from outside only by their greater number of antennae. The Panther saw service on all fronts from 1943 on.

Even though more than 6000 Panthers were built, it was never possible to equip all the Wehrmacht's battle tank units with them. Usually there was just one armored unit in each division equipped with them. The other units used the Battle Tank IV, and some, on account of a lack of battle tanks, were supplied with Assault Gun III.

As was customary for all tank types since 1940, this chassis was also used for special versions. The best-known of them are:
—Observation Tank "Panther",
—Recovery Tank "Recovery Panther",
—Pursuit Tank "Hunting Panther", with an 8.8 cm Pak L/71.

At the end of the war, great numbers fell into the hands of the victors. The French Army took over the majority of them and kept them into the Fifties. This too is surely a sign of its high quality.

This Panther type cannot be determined, since the last series of the first Type D tanks also bore cast turrets.

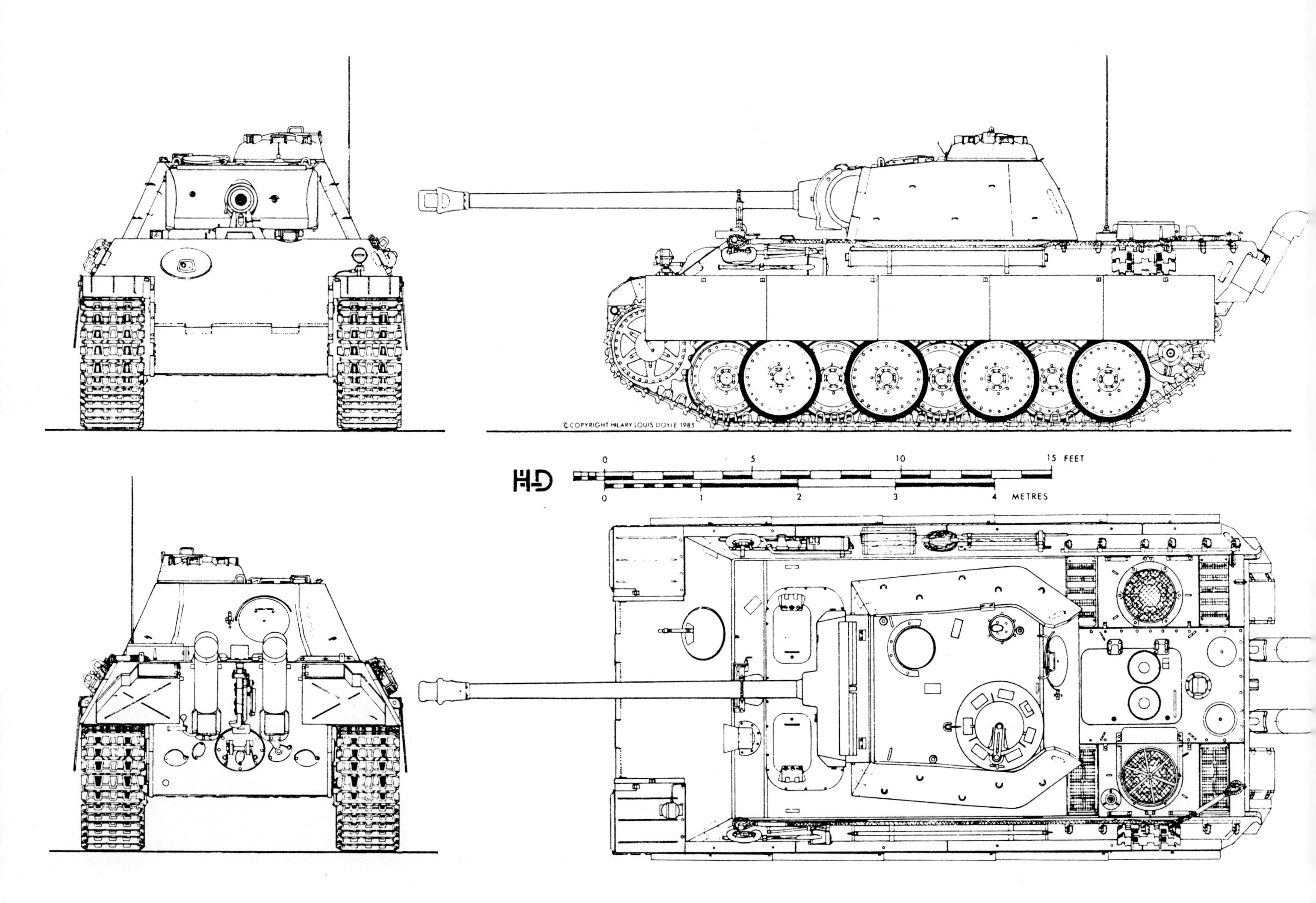

BATTLE TANK 'PANTHER' TYPE G.

Technical Data

Battle Tank V Type G

Crew	5 men
Overall dimensions	8860 x 3430 x 3000
Armor plate	Front 80 mm, sides and rear 40 mm
Armament	7.5 cm KwK 42 L/70, 3 machine guns (with anti-aircraft machine guns)
Ammunition	82 7.5 cm rounds, 4200 machine gun rounds
Gross weight	44.8 tons
Track width	850 mm
Ground pressure	0.90 kg/sq. cm
Motor	Maybach HL 230 P 30 12-cylinder
Horsepower	700 HP
Power to weight	15.4 HP
Fuel capacity	730 liters
Fuel consumption	412 liters/100 km (road)
Top speed	46 kph
Range (on road)	177 km
Wading ability	170 cm
Ground clearance	56 cm
Crossing ability	191 cm
Climbing ability	91 cm

Two type D command tanks, recognizable by the early cylindrical cupolas, the numbers on the turrets (R01 + the regimental commander's car, I01-that of the commander of the first unit of the regiment), and the umbrella antennae.

A Type A, as shown by the machine gun and the opened driver's visor on the bow.

TIGER I—Type E

The Battle Tank VI—Tiger I—Sd. Kfz. 181

Even though its was being planned since the spring of 1941—before the Russian campaign began—it can be regarded as an answer to the Russian T-34 and KW series tanks, as its development was speeded up considerably after their appearance. In addition to the number "VI" it was given the type name of "Tiger". While the Panther still belonged to the medium tank class, the Tiger, even before being armed, was already classed as a heavy tank. Its main armament was a variant of the renowned 8.8 cm anti-aircraft gun, with its velocity, high for those times, of 810 meters per second. Its crew consisted of five men.

It was not supplied to the already existing armored regiments. Planned as a concentration-point weapon, it had independent units—"heavy armored units"—created for it, to be sent into action by the High Command wherever needed. These units (501 through 510, plus a few with special units and the Waffen-SS) consisted of three companies each, with 14 Tigers each, plus three for the unit staff—a total of 45 Tigers.

Since its production came about earlier than that of the Panther, it saw action—though not exactly convincing action—at company strength at Leningrad at the end of 1942 and Rostov in January of 1943. The Heavy Armored Unit 501 was the first closed unit to go to North Africa (Tunis), even before the end of 1942, where it—often operating dispersed, along with Unit 504—caused considerable losses to the British and Americans until the surrender of May 1943. The first major attack—now along with the Panther—did not take place until the summer of 1943 in "Operation Citadel".

It was the equal of all opposing battle tanks and remained so until the war ended, despite several technical problems at the beginning and its unfavorable power-to-weight ratio (HP per ton) of only 10.6, along with armament and armor the decisive factors for evaluating a tank.

Aside from its further development into the Tiger II (King Tiger), for which the next chapter is reserved, there was officially only one (E) type. Yet this battle tank also showed many changes during the course of its production, including the elimination of the rear turret visor, a variety of commander's cupolas, more powerful motors, removal of the deep-wading (Snorkel etc.) equipment, and different types of road wheels. Only with the construction of the Tiger II was it given the additional designation "I" to avoid confusion. As with the Panther, there was also a command-car version of it. This too can be recognized from outside only by the additional radio equipment.

The following special versions can be noted:

—Assault Tank "Storm Tiger" with rocket launcher 61 (38 cm caliber), and
—Recovery Tank "Recovery Tiger".

In all, 1350 Tiger I were built. Some of them can still be seen as special exhibits in museums.

Its present-day legendary fame is based more on the Allied news reports of the time than on reality. To be sure, its gun was excellent and its armor was good, but its external mass was too great (unfavorable for railroad transport), its motor—even the more powerful type—was too weak for its great weight (its cross-country movements were sluggish and the motor often needed repairs because of being overstrained), its weight was too great for bridges and swampy terrain, and its design was not good for deflecting shots.

This photo is of interest because it shows the cast commander's cupola (Panther type) and the Zimmerite covering to keep off contact mines.

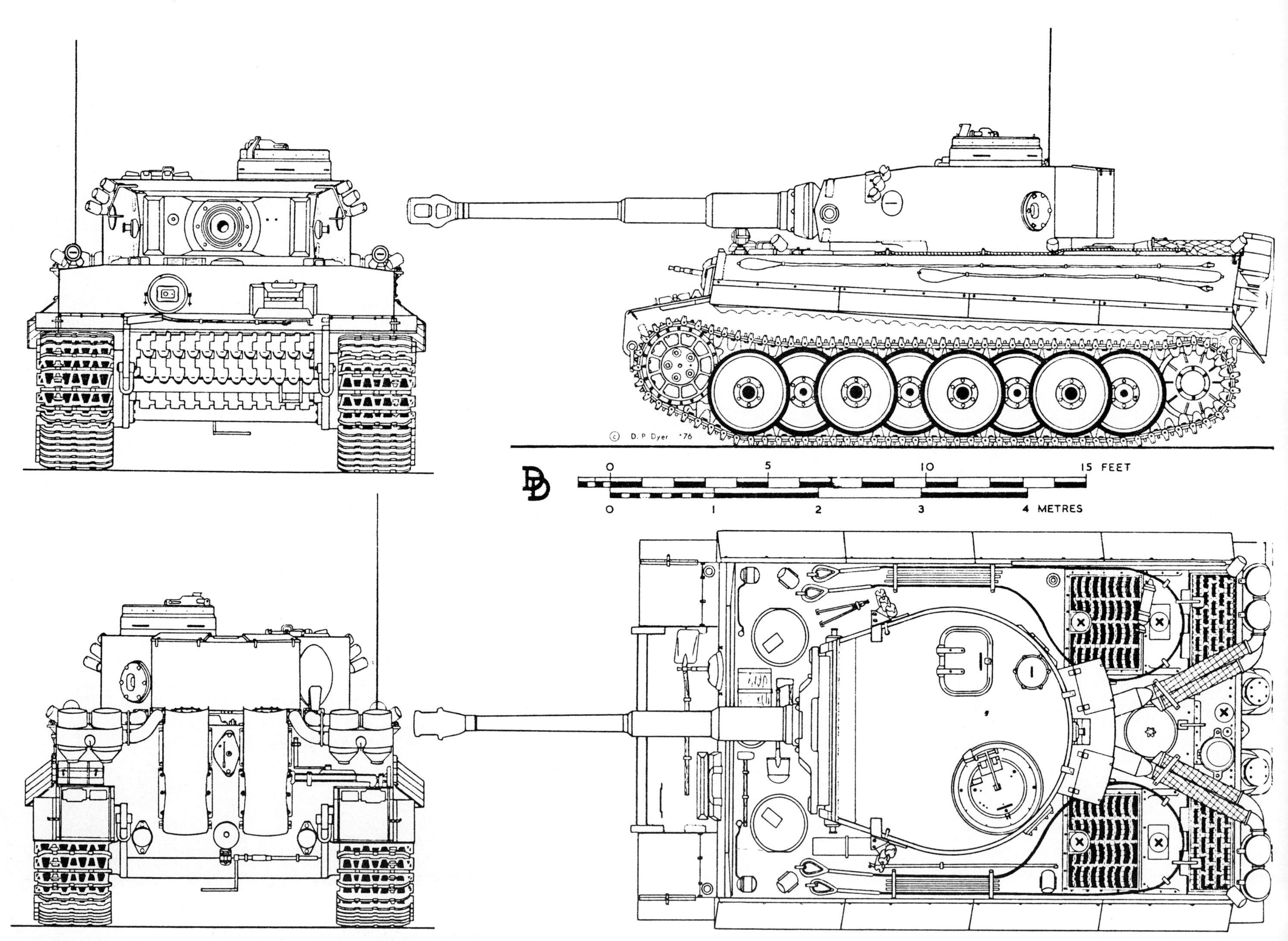
© D. P. Dyer '76
0
5
10
15 FEET
0
1
2
3
4 METRES

Technical Data

Battle Tank VI Tiger

Crew	5 men
Overall dimensions	8241 x 3705 x 2880
Armor plate	Front 100 mm, sides 60 mm, rear 80 mm
Armament	1 8.8 cm KwK 36 L/56, 2 machine guns
Ammunition	92 8.8 cm rounds, 4500 machine gun rounds
Gross weight	56.9 tons
Track width	25 mm
Ground pressure	1.00 kg/sq. cm
Motor	Maybach HL 230 P 45 12-cylinder
Horsepower	600 HP
Power to weight	10.6 HP
Fuel capacity	534 liters
Fuel consumption	over 500 liters/100 km (road)
Top speed	40 kpk (sustained speed 20 kph)
Range (on road)	approximately 100 km
Wading ability	120 cm (the first 500 tanks had deep-wading equipment allowing 396 cm.)
Ground clearance	43 cm
Crossing ability	180 cm
Climbing ability	79 cm

The Tiger I appeared with various camouflage paint styles: one-, two and (more rarely) three-color. It is seen above in two colors (dark green on tan) and below in three (dark yellow, red-brown and olive green).

234

The Battle Tank VI—Tiger II-Sd. Kfz. 182

As early as the spring of 1943, the results of the first experience with the Tiger resulted in an appeal for a more shot-deflecting form. Thus began the development of a new Battle Tank VI, which went into production early in 1944 and was designated "Tiger II". But since it was more heavily armored and carried a longer 8.8 cm KwK (L/71 instead of the previous L/56), its weight increased to 70 tons. The result was a new set of running gear, though a new motor would have been more important. But since that did not transpire, the Tiger II turned out to be even more sluggish on the battlefield than the earlier Battle Tank VI, now called "Tiger I". Thus despite its externally improved form it was no supertank—as has often been claimed—as its very long gun was made unstable by its length and fired less accurately than the L/56, and was even "unhandier" for fighting in towns and forests.

It too was delivered to independent heavy armored units, and there was officially one (B) type; yet its production runs showed variations. The first fifty Tiger II had the turret of a battle tank originally planned as a Porsche Tiger, which was later rebuilt into the "Ferdinand" pursuit tank, later called the "Elefant". Only from the 51st Tiger II on did all the rest have the "production turret" made by Krupp and equipped with a different gun shield. Here too, the command tanks can be recognized by the multiple antennae. In all, 487 of them were built.

It saw action mainly in the West. There the Allies named it "King Tiger", which was then picked up by the Germans (never officially!) and is the customary name used for it today.

The following special versions of it existed:

—Pursuit tank "Jagdtiger" (Hunting Tiger), and

—Recovery Tank "Recovery Tiger II".

Of these, the Hunting Tiger is to this day the heaviest pursuit tank in the world, usually being armed with a 12.8 cm antitank gun.

A Tiger II with the production turret made by Krupp.

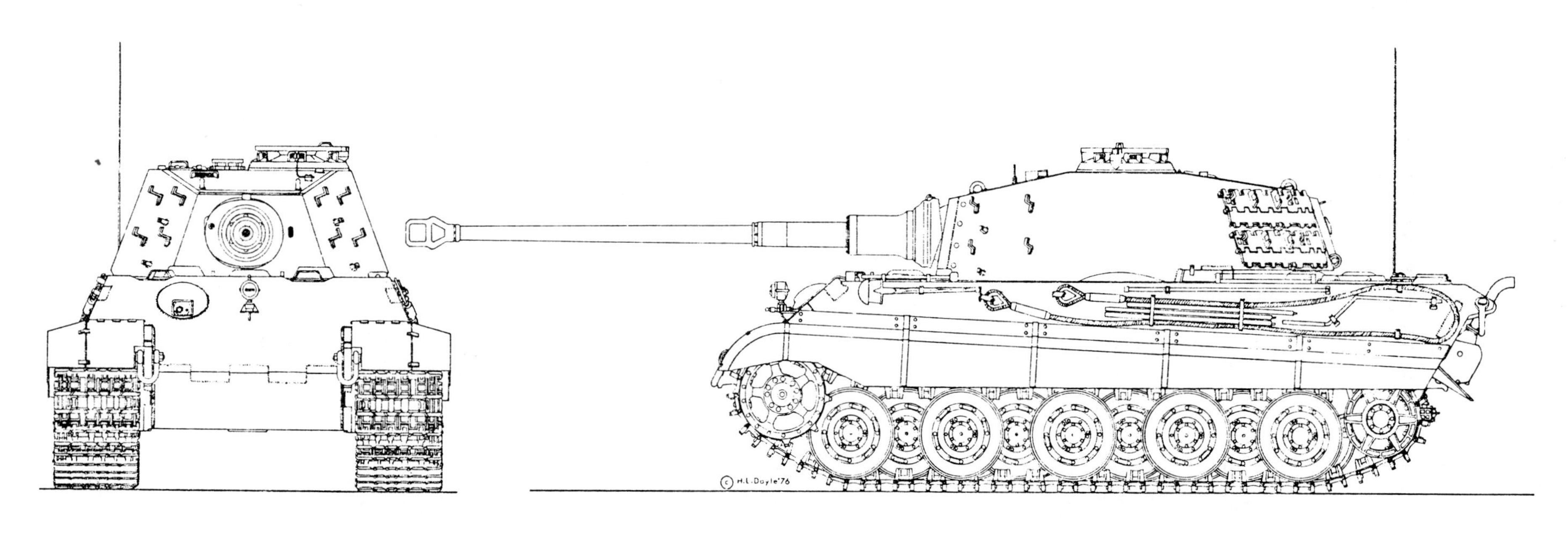

BATTLE TANK 'TIGER' TYPE B.

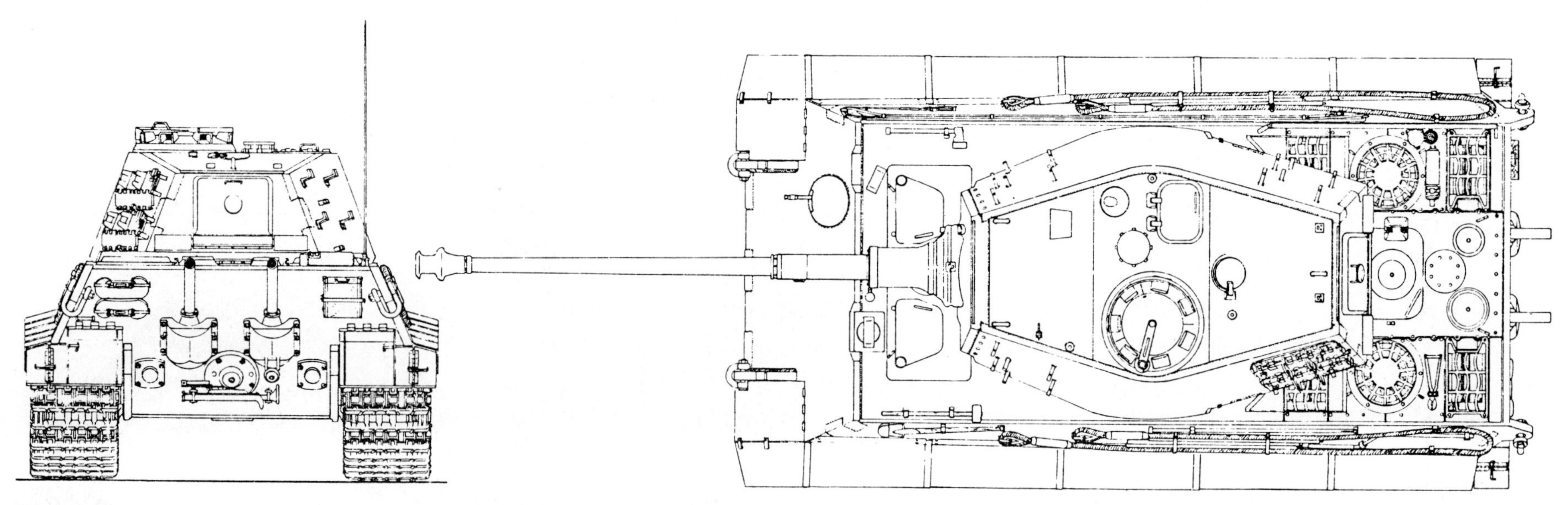

Technical Data

Battle Tank VI Tiger II (Production Turret)

Crew	5 men
Overall dimensions	10,260 x 3750 x 3090
Armor plate	Front 150 mm, sides and rear 80 mm
Armament	1 8.8 cm KwK 43 L/71, 3 machine guns (with anti-aircraft machine guns)
Ammunition	84 8.8 cm rounds, 5850 machine gun rounds
Gross weight	70 tons
Track width	800 mm
Ground pressure	1.07 kg/sq. cm
Motor	Maybach HL 230 P 45 12-cylinder
Horsepower	600 HP
Power to weight	8.6 HP
Fuel capacity	860 liters
Fuel consumption	782 liters/100 km (road)
Top speed	38 kph
Range (on road)	160 km
Wading ability	160 cm
Ground clearance	50 cm
Crossing ability	250 mm
Climbing ability	85 mm

The first fifty Tiger II tanks used the narrow Porsche turret.

This photo shows clearly the shot-deflecting shape, the long gun and the very wide tracks of the King Tiger.

201
THIJS POSTMA

The Battle Tank Maus

The idea of building heavily armored and large-caliber penetration tanks had been around since World War I, and had its adherents in Germany as elsewhere—and not the least of them was Hitler himself. Thus there arose—what with the impression made by the numerous Russian heavy tanks—along with many other projects, most of which never got farther than the drawing board, the super-heavy "MAUS" battle tank designed by Dr. Porsche.

At the end of 1943 the first prototype was finished by Alkett. It had front armor of 240 mm, a 12.8 cm KwK with a coaxial 7.5 cm KwK L/44, but only 1080 horsepower to move a gross weight of 188 tons. This resulted, particularly with the inevitably unfavorable power-to-weight ratio of 5.7 HP per ton and its huge dimensions, to incredible difficulties. The "Maus" had exceeded the limits of tank construction!

A second "Maus" was, in fact, built, and at war's end parts for three others were found, but these battle tanks never went into battle. Weights over fifty tons (bridges!), too-great mass (transport difficulties), and too-unfavorable power-to-weight ratios (powerless!) were always, and are still today, wrong directions for tank construction.

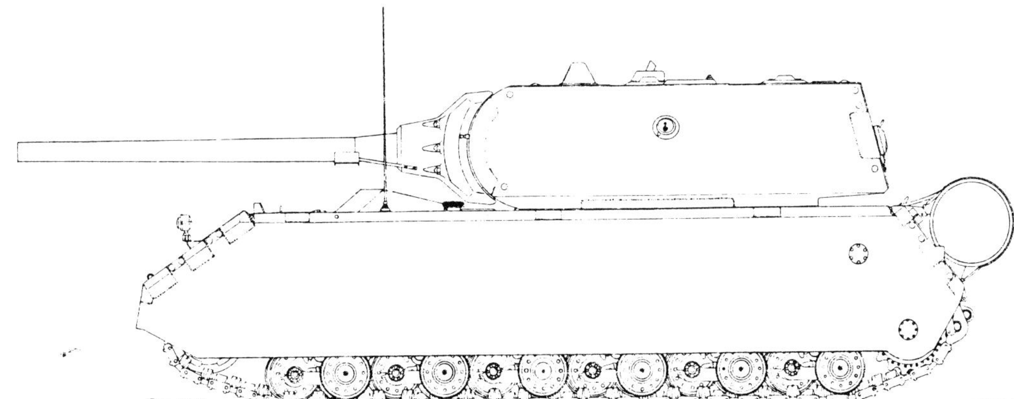

Technical Data

Battle Tank "Maus"

Crew	6 men
Overall dimensions	10080 x 3670 x 3630
Armor plate	Front 240 mm, sides and rear 200 mm
Armament	1 12.8 cm KwK 44 L/68, 1 7.5 cm KwK 44
Ammunition	Unknown
Gross weight	188 tons
Track width	1100 mm
Ground pressure	1.31 kg/sq. cm
Motor	Daimler-Benz MB 509
Horsepower	1080 HP
Power to weight	5.6 HP
Fuel capacity	2700 liters
Fuel consumption	1400 liters/100 km (road)
Top speed	20 kph
Range (on road)	190 km
Wading ability	200 cm, 800 when diving
Ground clearance	54 cm
Crossing ability	450 cm
Climbing ability	72 cm

Comparison Data

Type	Crew	Main Weapon	Armor thickness	Tons	HP/ Ton	Ground pressure	Road Range (km)
Pz. IB	2	MG	13	5.9	17	.42	180
Pz. IIF	3	2 cm	35	9.5	14.7	.66	150
Pz 35	4	3.7 cm	25	10.5	11.4	.52	190
Pz 38E	4	3.7 cm	25	10.5	14.3	.57	250
Pz. III H	5	5 cm L/42	30	21.6	12.2	.99	140
Pz. III M	5	5 cm L/60	70	22.3	12.3	.94	155
Pz. IV E	5	7.5 cm L/24	30	21	12.6	.79	230
Pz. IV H	5	7.5 cm L/48	80	25	10.6	.89	180
Pz. V G	5	7.5 cm L/70	80	44.8	15.4	.9	177
Pz. VI/I	5	8.8 cm L/56	100	56.9	10.6	1.09	100
Pz. VI/II	5	8.8 cm L/71	150	70	8.6	1.07	110
Maus	6	12.8 cm	240	188	5.6	1.31	190
T-34/76A	4	7.62 cm	45	26.3	19	.6	450
Stalin I(122)	4	12.2 cm	120	45	11.5	.77	240
Leo 2	4	12 cm	Layered	54	27.3	.83	600
T-72	3	12.5 cm	Layered	41	21.8	.9	500

This chart of the most important data of the tanks discussed in this book (other than the "Neubaufahrzeug") show very clearly the tendencies of tank construction at that time, as well as strengths and weaknesses of the individual types. From this data, judgments of fighting power and mobility can easily be made, the latter in particular from the three columns: Power-to-weight ratio, ground pressure and range. A power-to-weight ratio under 10 indicates sluggish motion off the road, a ground pressure over 0.90 requires a firm roadbed, and a road range under 200 (which means much less across country!) indicates the necessity of maintaining a fuel supply very close to these tanks.

The comparisons with the two most important Russian tanks of World War II and the two presently most modern European battle tanks, the German Leopard 2 and the Russian T-72, are also interesting.

KEY TO THE COLOR DRAWING AT RIGHT:

1. Three fuel tanks (on both sides)
2. Exhaust
3. Fuel fillers
4. Ventilator
5. Air inlet grille
6. Air cooler for the motor
7. Armored engine cover
8. Turret (production type)
9. Commander's seat
10. Commander's cupola
11. Gunner's seat (left), loader's seat (right)
12. 8.8 cm KwK 1/71
13. Coaxial MG 34
14. Ventilator
15. Rifling in gun barrel
16. Driver's visor
17. Shock absorbers
18. Driver's seat
19. Headrest for the radioman (machine-gunner)
20. Gearbox
21. Round shield for MG 24
22. Six 8.8 cm shells
23. Radioman's seat